NI-FREE TI-BASED SHAPE MEMORY ALLOYS

NI-FREE TI-BASED SHAPE MEMORY ALLOYS

HEE YOUNG KIM

SHUICHI MIYAZAKI

ELSEVIER

Butterworth-Heinemann
An imprint of Elsevier

Butterworth-Heinemann is an imprint of Elsevier
The Boulevard, Langford Lane, Kidlington, Oxford OX5 1GB, United Kingdom
50 Hampshire Street, 5th Floor, Cambridge, MA 02139, United States

Library of Congress Cataloging-in-Publication Data
A catalog record for this book is available from the Library of Congress

British Library Cataloguing-in-Publication Data
A catalogue record for this book is available from the British Library

ISBN: 978-0-12-809401-3

For information on all Butterworth-Heinemann publications
visit our website at https://www.elsevier.com/books-and-journals

Publisher: Matthew Deans
Acquisition Editor: Christina Gilfford
Editorial Project Manager: Gabriela D. Capille
Production Project Manager: R.Vijay Bharath
Cover Designer: Mark Rogers

Typeset by SPi Global, India

CONTENTS

PREFACE

Since the discovery of shape memory effect in Ti-Ni in the 1960s and the achievement of superelasticity in the 1980s, Ti-Ni-based shape memory alloys have attracted increasing attention and interest. Owing to their outstanding shape memory and superelastic properties, Ti-Ni-based shape memory alloys have become commercially popular and have been successfully used in many industrial fields. Ti-Ni-based shape memory alloys in recent years have emerged into biomedical fields and have been used as health-care and biomedical devices, e.g., eyeglass frames, orthodontic arch wires, guide wires, and self-expanding stents. Ti-Ni-based shape memory alloys have also been considered as attractive materials for biomedical implants. However, there have been long-standing concerns regarding the risk of Ni allergy and hypersensitivity for long-term use because of a high amount of Ni content in Ti-Ni-based shape memory alloys. Over the last decade, β-type Ti-based alloys have been extensively studied as promising candidates for Ni-free biomedical shape memory alloys. There have been impressive progresses in understanding the basic characteristics of the martensitic transformation and shape memory behavior of the β-type Ti-based alloys, and many Ti-based shape memory alloys have been developed. It is confidently expected that superelastic Ti-based alloys will further expand the applications of shape memory alloys in biomedical field. This book comprises seven chapters designed to cover all fundamental issues of β-type Ti-based shape memory alloys. Chapter 1 covers the fundamental characteristics of the martensitic transformation in Ti-based alloys, including transformation temperature, crystal structure, habit plane, self-accommodation morphology, and ω transformation. Chapter 2 provides the overview of the shape memory and superelastic properties of Ti-based alloys. Chapter 3 discusses the effects of interstitial alloying elements on microstructure, mechanical properties, and transformation behavior. Chapter 4 focuses on thermomechanical treatment and microstructure control of Ti-based shape memory alloys. Relationships between microstructure parameters and shape memory properties are discussed. Chapter 5 covers the issues of unique properties of metastable β Ti alloys. The roles of the martensitic transformation on nonlinear elasticity and Invar-like thermal expansion behavior in gum metal, {332} twinning, and low Young's modulus of metastable β Ti alloys are discussed. Chapter 6 provides the overview of the cytocompatibility and electrochemical characteristics of Ti-based shape memory alloys. Chapter 7 introduces the fabrication processes of commercial Ti-Ni-based shape memory alloys and Ni-free Ti-based shape memory alloys. We hope this book contributes to further understanding of the martensitic transformation and shape memory effect of β Ti alloys and providing useful information for researchers, engineers, and graduate students working in materials science, biomedical materials, metallurgy, and light alloys and professionals in the titanium industry. Finally, we express special thanks to the staffs at ELSEVIER for their continuous support and assistance.

CHAPTER 1

Martensitic Transformation Characteristics

Contents

1.1 INTRODUCTION

Understanding of martensitic transformation is essential to develop shape memory alloys and to improve their properties since the shape memory effect is based on the reversible thermoelastic martensitic transformation. Martensitic transformation characteristics of β-type Ti-based alloys such as transformation temperature and habit plane have been investigated since the 1950s [1–6]. Duwez [1] first investigated the Mo content dependence of the martensitic transformation in Ti-Mo binary alloys using the thermal-arrest technique and reported that the martensitic transformation temperature decreases progressively with increasing Mo content. Liu and

Ni-free Ti-based Shape Memory Alloys
https://doi.org/10.1016/B978-0-12-809401-3.00001-X

Margolin [2] investigated the composition dependence of martensite structure and habit plane in Ti-Mn alloys. In the early 1960s, Kaneko and Huang [7] systematically investigated the martensitic transformation temperature as a function of alloy composition in several binary titanium alloy systems, e.g., Ti-Nb, Ti-V, Ti-Mo, Ti-Al, and Ti-Sn. Since then, there have been continued efforts to understand the nature of martensitic transformation in Ti alloys. As a result, many aspects of the martensitic transformation in Ti alloys, such as transformation temperature [8–17], crystal structure [8, 16, 18–29], habit planes, and substructure of martensite [30–39], have been clarified. However, most of previous studies focused on the dilute alloys with a low content of alloying element. There have been limited studies on martensitic transformation in β-type Ti-based alloys that exhibit shape memory effect and superelastic behavior. In β-type Ti-based alloys, the $\beta \rightarrow \omega$ transformation is in competition with the martensitic transformation and has attracted considerable attention. It has become clear that the ω phase greatly affects the martensitic transformation behavior and shape memory properties of β-type Ti-based alloys [40–42]. In this chapter, the fundamental characteristics of the martensitic transformation in Ti-based alloys are overviewed. The effect of composition on transformation temperature and crystal structure in Ti-Nb, Ti-Mo, and Ti-Ta binary alloys, which are most attractive alloys for shape memory applications, is particularly reviewed. Some important aspects of martensite structures such as habit plane, interface, and self-accommodation morphology are also summarized. The effect of ω phase on the martensitic transformation behavior and superelastic properties is also discussed.

1.2 CRYSTAL STRUCTURE AND MARTENSITIC TRANSFORMATION TEMPERATURE

1.2.1 Crystal Structure of Parent and Martensite Phases

The β phase, i.e., high-temperature phase or parent phase, has a bcc structure, while the martensite phase has a hexagonal structure (α' phase) or an orthorhombic structure (α'' phase). Fig. 1.1 shows the crystal structures of β, α', and α'' phases and their lattice correspondence. Four unit cells of β phase, where the lattice parameter is a_0, are drawn with thin solid lines. A face-centered tetragonal cell ($b_0 = c_0 = \sqrt{2}a_0$) in the β-phase cells is transformed into an orthorhombic structure with lattice parameters of a, b, and c ($a < c < b$) through the martensitic transformation $\beta \rightarrow \alpha''$. This cubic-to-orthorhombic martensitic transformation also occurs in several shape

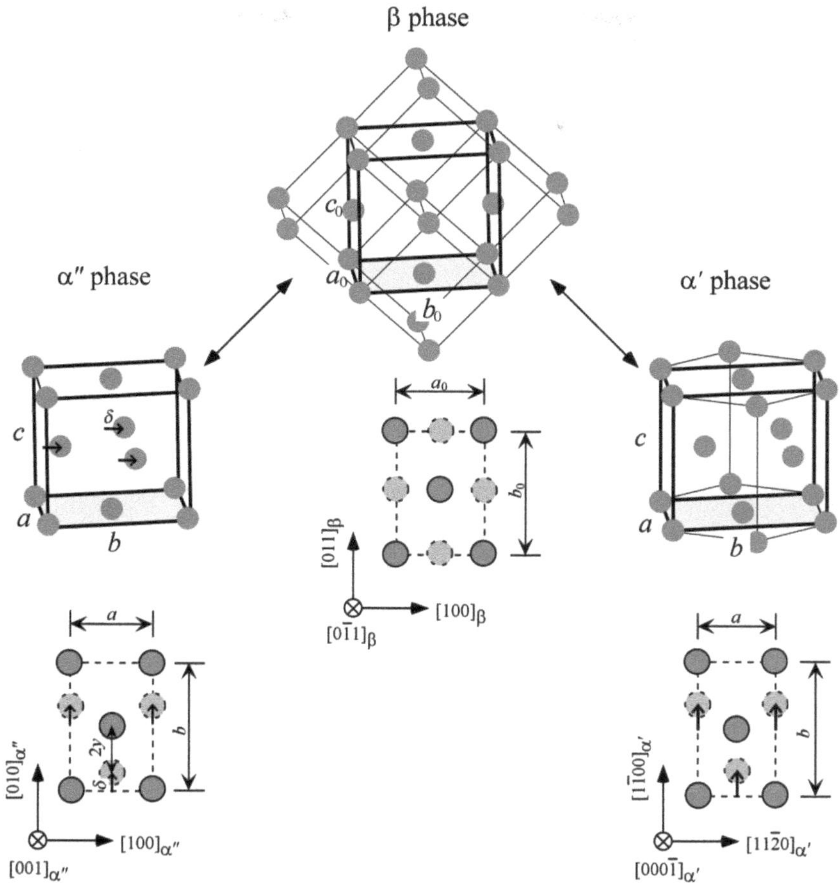

Fig. 1.1 Crystal structures of β, α′, and α″ phases and their lattice correspondence.

memory alloys such as Au-Cd [43], Ti-Pt [44], Ti-Ni-Cu [45], and Cu-Al-Ni [46]. The hexagonal structure can be treated as an orthorhombic structure where the magnitude of b in the orthorhombic cell corresponds to $\sqrt{3}a$.

Fig. 1.2 shows the Nb content dependence of lattice parameters of α′ and α″ martensite phases (a, b, and c) and β phase (a_0) at room temperature in the binary Ti-Nb alloys taken from the literature [5, 8, 9, 12, 16, 19, 20, 22–25, 27, 28]. In the Ti-Nb alloys, the martensite phase changes from α′ to α″ at around 7 at.% Nb. For the α″ phase, the lattice parameters changes almost linearly with Nb content: b decreases, while a increases with increasing Nb content. The change of c is small but increases slightly with increasing Nb content until around 7 at.% and decreases slightly with further increasing

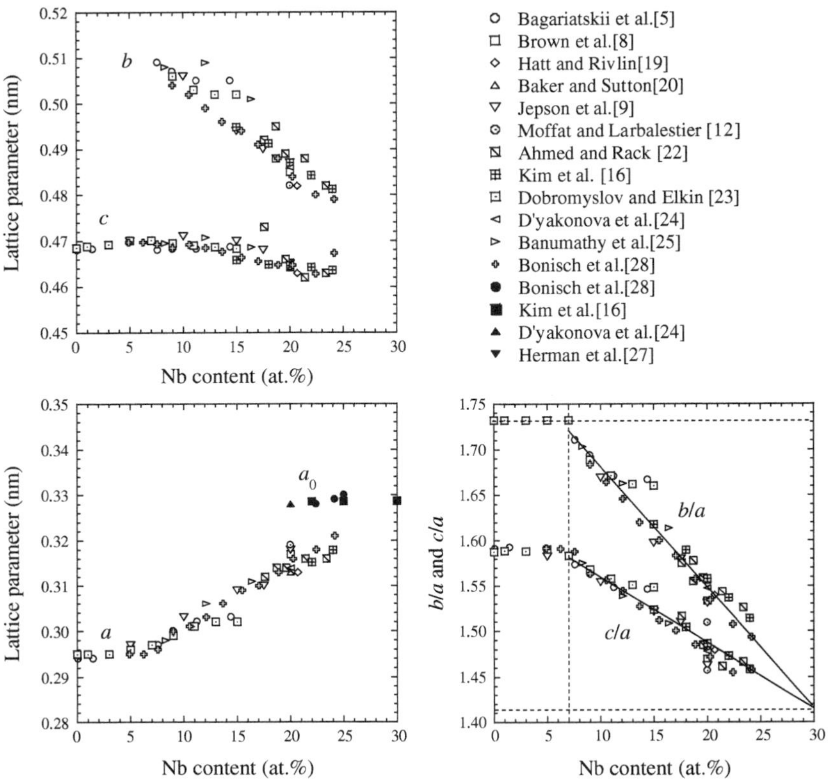

Fig. 1.2 Nb content dependence of lattice parameters of martensite phases (*a*, *b*, and *c*) and β phase (a_0) for the binary Ti-Nb alloys.

Nb content. Fig. 1.2 also shows the Nb content dependence of b/a and c/a. For the α′ martensite phase, c/a and b/a are not dependent on the Nb content. The values of b/a (1.73) and c/a (1.59) are similar to those of equilibrium α phase. On the other hand, b/a and c/a decrease linearly with increasing Nb content for the α″ martensite phase. This indicates that the crystal structure of the α″ martensite phase becomes closer to a bcc structure. The extrapolation predicts that both b/a and c/a become $\sqrt{2}$ at around 30 at. % Nb, implying that the orthorhombicity of α″ martensite phase disappears. These results support the hypothesis that the α″ orthorhombic martensite phase is a compromise structure or an intermediate phase between the β phase with a bcc structure and the α′ phase with a hexagonal structure [47].

Fig. 1.3 shows the Ta content dependence of lattice parameters of α′ and α″ martensite phases in the binary Ti-Ta alloys [31, 48–52]. Although literature data are somewhat scattered, the trends of lattice parameters of

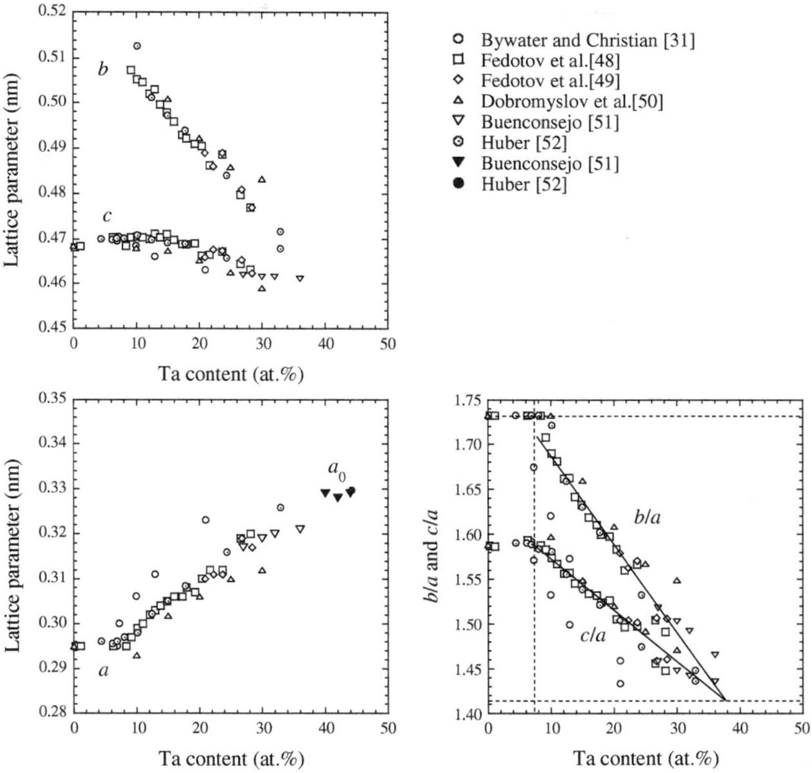

Fig. 1.3 Ta content dependence of lattice parameters of martensite phases (*a*, *b*, and *c*) and β phase (a_0) for the binary Ti-Ta alloys.

martensite phases with the increase of Ta content in Ti-Ta alloys are similar to those of Ti-Nb alloys: the α′ phase appears as a martensite phase until the Ta content reaches about 7 at.%, but the β phase transforms to the α″ phase at higher concentrations. The changes in *c*/*a* and *b*/*a* with respect to the Ta content is slightly mild when compared with Ti-Nb alloys shown in Fig. 1.2, and the upper limit of α″ martensite phase is around 40 at.% Ta.

There have been limited reports characterizing crystal structure of martensite phases in other binary alloy systems such as Ti-Mo alloys [18, 21, 23, 26, 30, 53] as shown in Fig. 1.4. In the Ti-Mo alloys, the crystal structure of martensite phase changes from hexagonal to orthorhombic at a much lower concentration of around 2 at.% as compared with Ti-Nb and Ti-Ta alloys. There are few reliable results ever reported in Ti-Mo alloys. Although the extrapolation of *c*/*a* and *b*/*a* predicts that the upper limit of α″ martensite phase is about 11 at.% Mo, the β → α″ martensitic transformation has been

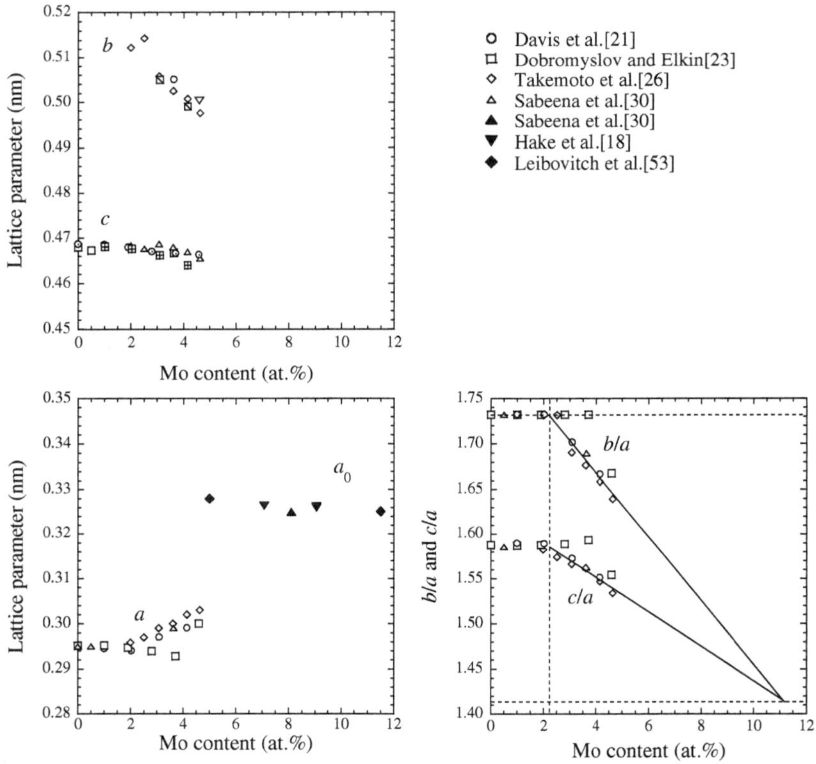

Fig. 1.4 Mo content dependence of lattice parameters of martensite phases (*a*, *b*, and *c*) and β phase (a_0) for the binary Ti-Mo alloys.

observed in a limited composition range up to about 5 at.% Mo. This is due to the $\beta \rightarrow \omega$ transformation tends to occur prior to the $\beta \rightarrow \alpha''$ martensitic transformation. The effect of the ω phase on the $\beta \rightarrow \alpha''$ martensitic transformation and shape memory effect will be described in more detail later.

In addition to the lattice deformation or lattice distortion as explained above, the martensitic transformation results in a rearrangement of atomic positions in a unit cell, known as shuffling. As shown in Fig. 1.1, the unit cell of the α″ martensite structure is considered as a c-centered orthorhombic structure with space group of *Cmcm*. There have been some controversies on the crystal structure of α″ phase [22, 33, 38] based on the antiphase boundary (APB) like morphology and appearance of 00*l* (*l*=odd) superlattice reflections in the diffraction patterns. In order to explain these APB-like morphologies and diffraction patterns, Pak et al. [33] and Ahmed and Rack [22] claimed that the α″ phase has an ordered structure. However, the α″

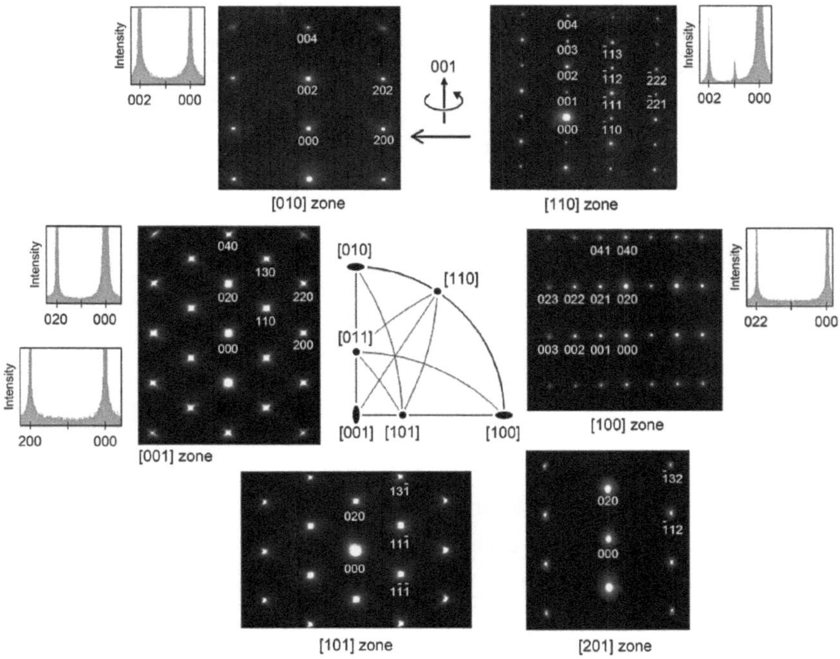

Fig. 1.5 Selected area diffraction patterns taken from α″ martensite in a Ti-23Nb-3Al alloy.

martensite has been confirmed to be disordered similarly to the β phase [38]. Fig. 1.5 shows a series of diffraction patterns taken from α″ martensite in Ti-23Nb-3Al [38]. The 00*l* (*l*=odd numbers) superlattice reflections are seen in the diffraction patterns from the [100] and [110] zone axes; however, they are absent in the [010] diffraction pattern. This result indicates that the 00*l* (*l*=odd) superlattice reflections are not due to the intrinsic diffraction but due to the multiple diffraction. It is noted that, in most of nonferrous shape memory alloys, both parent and martensite phases have ordered structure, e.g., B2 → B19′ in TiNi [54]; B2 → B19 in TiPd, TiAu, and TiPt [55]; $L2_1$ → 18R in Zn_2AuCu [56]; $L2_1$ → $L1_0$ in Ni_2MnGa [57]; and B2 → 3R (2M) or 7R (14M) in NiAl [58]. Besides β-type Ti-based alloys, limited shape memory alloys have disordered crystal structures, e.g., FeMnAl (A2 bcc → A1 fcc) [59].

When comparing the positions of atoms of the β phase and the α″ phase in Fig. 1.1, the atoms in the (002) plane are shifted along the [010] direction in the orthorhombic unit cell upon the martensitic transformation. The magnitude of shuffle δ is $1/2-2y$, where y identifies the atomic positions on the (002) plane

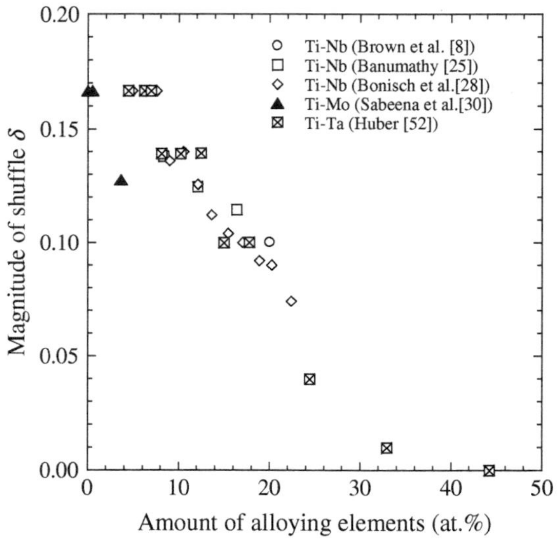

Fig. 1.6 Magnitude of shuffle as a function of alloying content for the Ti-Nb, Ti-Ta, and Ti-Mo alloys.

along the [010] direction. The magnitude of shuffle is dependent on the composition. It is noted that the magnitude of shuffle in the α' structure is 1/6, and its range in the α'' structure is between 0 and 1/6. Fig. 1.6 shows the magnitude of shuffle with respect to the amount of alloying elements in the Ti-Nb, Ti-Ta, and Ti-Mo alloys [8, 25, 28, 30, 52]. It is seen that the magnitude of shuffle decreases as the amount of alloying elements increases. Fig. 1.7 shows a good correlation between the magnitude of shuffle and the b/a value; however, for valid comparison and analysis, more accurate experimental data on the crystal structures and atomic positions are required. According to the lattice correspondence, the shuffling mode is expressed as $\{110\}\langle 1\overline{1}0\rangle$ in the β phase, i.e., shuffling of parallel adjacent $\{110\}$ planes to $\langle 1\overline{1}0\rangle$ directions. It has been well acknowledged that the shear modulus C' $(=(C_{11}-C_{12})/2)$ corresponds to the stiffness associated with a shear $\langle 1\overline{1}0\rangle$ directions and therefore being a key parameter in determining the stability of the bcc structure [60–63]. Table 1.1 compares the values of shear moduli C', C_{44}, and anisotropy ratio A $(=C_{44}/C')$ of the Ti-Nb-based shape memory alloys [64–66] with those of the other representative shape memory alloys with bcc-based structures [67–72]. It is seen that Ti-Nb-based shape memory alloys exhibit low C' in the range of 10–13 GPa as expected, which is comparable with other shape memory alloys. Also, it is noted that the values of C_{44} of Ti-Nb-based shape memory alloys are lower than those of other shape memory alloys.

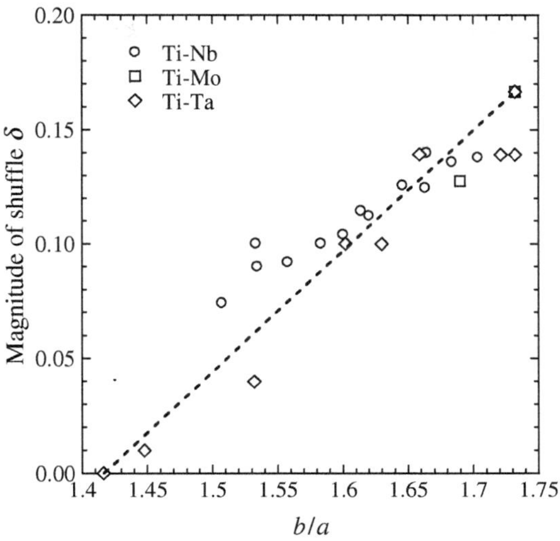

Fig. 1.7 Relationship between the magnitude of shuffle and the *b/a* ratio for the Ti-Nb, Ti-Ta, and Ti-Mo alloys.

Table 1.1 Comparison of stiffness coefficients of the Ti-Nb-based shape memory alloys with those of other bcc-based shape memory alloys

Alloy	C' (GPa)	C_{44} (GPa)	*A*	Reference
Ti-40Nb (wt%)	12.2	33.1	2.71	[64]
Ti-33Nb-4Sn (wt%)	12.6	22.4	1.78	[65]
Ti-24Nb-4Zr-8Sn (wt%)	10.6	35.9	3.40	[66]
Ti-Ni	17–19	35–39	2	[67–69]
Cu-Zn-Al	5.8	86	15	[70]
Au-Cd	3–3.8	42	11–14	[71]
Ni-Al	14.6	132	9	[72]

1.2.2 Transformation Temperature

Fig. 1.8 shows the martensitic transformation start (M_s) temperature with respect to the amount of alloying elements in the Ti-Nb, Ti-Ta, and Ti-Mo alloys [3, 7, 8, 9, 11, 15, 16, 51]. Compared with other alloys, the results on Ti-Nb alloys have been reported most frequently although there are some variations. The M_s temperature for both $\beta \rightarrow \alpha'$ and $\beta \rightarrow \alpha''$ transformation in Ti-Nb alloys decreases almost linearly with increasing Nb content. It is seen that the decreasing slope in the $\beta \rightarrow \alpha''$ transformation is steeper than that in the $\beta \rightarrow \alpha'$ transformation. The M_s temperature for $\beta \rightarrow \alpha''$ transformation in Ti-Nb alloys decreases to about room temperature

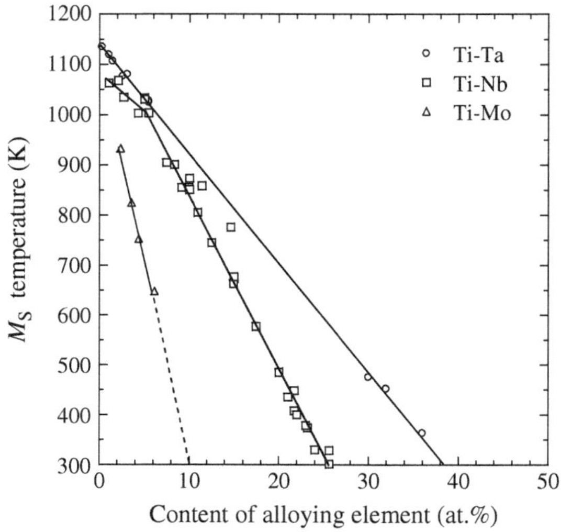

Fig. 1.8 Martensitic transformation start temperature as a function of alloying content for the Ti-Nb, Ti-Ta, and Ti-Mo alloys.

when the Nb content is about 25 at.%. In the Ti-Ta alloys, the M_s temperature decreases slowly with increasing Ta content, and the slope is much gentler than that in the Ti-Nb alloys. In contrast to the Ti-Nb and Ti-Ta alloys, the M_s temperatures in Ti-Mo alloys have been reported only at a high-temperature region, i.e., in the alloys with a small amount of Mo. The decreasing slope in Ti-Mo alloys is much steeper than those in the Ti-Nb and Ti-Ta alloys. The linear approximation formula of the M_s temperature in the $\beta \rightarrow \alpha''$ transformation for each alloy system is expressed as follows:

$$M_s(\mathrm{K}) = 1207 - 35.7X_{\mathrm{Nb}}$$

$$M_s(\mathrm{K}) = 1142 - 21.9X_{\mathrm{Ta}}$$

$$M_s(\mathrm{K}) = 1098 - 75.4X_{\mathrm{Mo}}$$

Fig. 1.9 shows the martensitic transformation start (M_s) temperature with respect to the amount of alloying elements of Al, Sn, and Zr [7, 10], which are major alloying elements in Ti-based shape memory alloys. The M_s temperature of Ti-Al alloys exhibits a slightly increasing tendency with increasing Al content. Although the slopes are not large, both Sn and Zr decrease the M_s temperature; with a slope of $-6\,\mathrm{K}/1$ at.% Sn in the Ti-Sn alloys and $-10\,\mathrm{K}/1$ at.% Zr in the Ti-Zr alloys. It should be noted that, in the Ti-Al, Ti-Sn, and Ti-Zr alloys, these M_s temperatures are associated with the $\beta \rightarrow \alpha'$ martensitic transformation. Fig. 1.10 shows the relationship

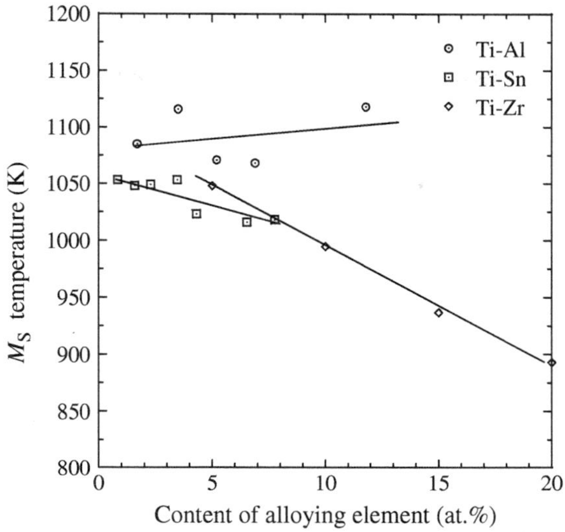

Fig. 1.9 Martensitic transformation start temperature as a function of alloying content for the Ti-Al, Ti-Sn, and Ti-Zr alloys.

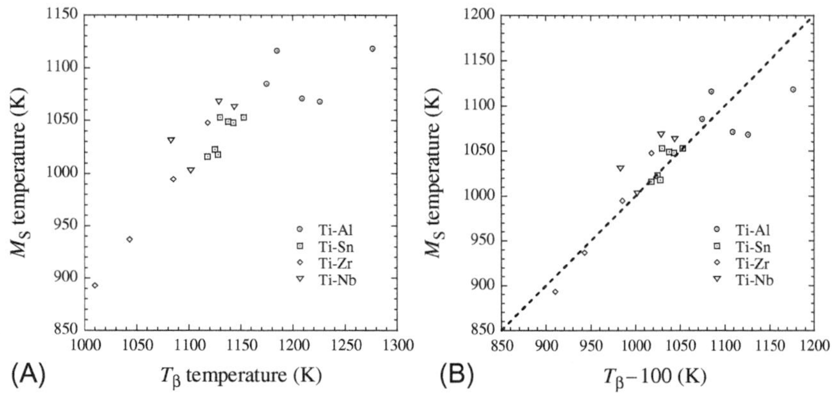

Fig. 1.10 Relationship between the M_s temperatures of the $\beta \rightarrow \alpha'$ martensitic transformation and the β transus temperature for Ti-Nb, Ti-Al, Ti-Sn, and Ti-Zr alloys. (A) M_s temperature versus T_β temperature and (B) M_s temperature versus $T_\beta - 100$ K.

between the M_s temperatures of the $\beta \rightarrow \alpha'$ martensitic transformation and the β transus temperature T_β, indicating that α-stabilizing alloying elements raise the M_s temperature but β-stabilizing alloying elements lower the M_s temperature of the $\beta \rightarrow \alpha'$ martensitic transformation. It can be roughly

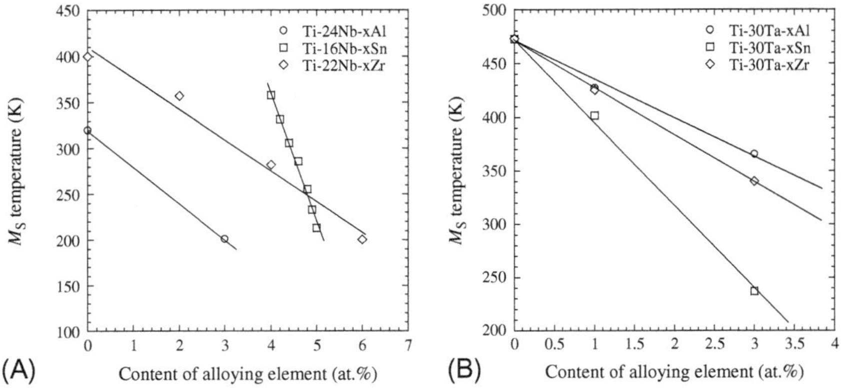

Fig. 1.11 Effect of the Al, Sn, and Zr addition on the martensitic transformation start temperature for (A) Ti-Nb and (B) Ti-Ta-based alloys.

estimated that the M_s temperature is 100 K less than the T_β of the alloy, i.e., $M_s(K) = T_\beta - 100$.

Controversial results have been reported on the effect of Al on the M_s temperature of the $\beta \rightarrow \alpha''$ martensitic transformation. As shown in Fig. 1.11A and B, the addition of Al in a Ti-24Nb alloy or a Ti-30Ta alloy obviously decreases the M_s temperature of the $\beta \rightarrow \alpha''$ martensitic transformation [73, 74], implying that Al suppresses the $\beta \rightarrow \alpha''$ martensitic transformation although it is an α-stabilizing alloying element. The slope is -40 K/1 at.% Al for the Ti-24Nb alloy and -35 K/1 at.% Al for the Ti-30Ta alloy. The addition of Zr and Sn also decreases the M_s temperature of the $\beta \rightarrow \alpha''$ martensitic transformation [75, 76]. It is also noted that the impact on the M_s temperature of the $\beta \rightarrow \alpha''$ martensitic transformation is much stronger when compared with the $\beta \rightarrow \alpha'$ martensitic transformation. The addition of Sn decreases the M_s temperature in the Ti-30Ta alloy with a slope of -79 K/1 at.%. In the Ti-16Nb-Sn alloys, the impact of Sn on the M_s temperature is much stronger: the M_s temperature decreases with a slope of -140 K/1 at.%, indicating that the influence of alloying elements on the M_s temperature is different depending on the alloy system.

Fig. 1.12 shows the effect of Nb content and Al content on the reverse transformation peak (A^*) temperature of Ti-18Zr-Nb-Al alloys [77]. The influence of Nb on decreasing the transformation temperature is much stronger than Al: the reverse transformation peak temperature decreases with a slope of -60 K/ at.% Nb and -25 K/at.% Al, respectively.

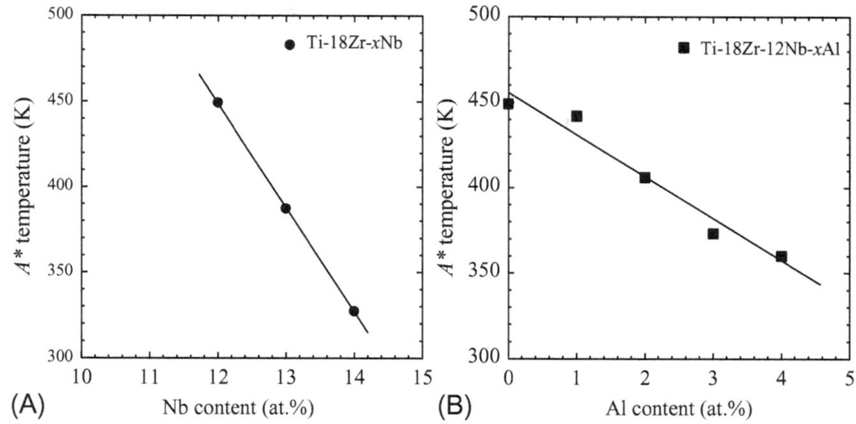

Fig. 1.12 Effect of the Nb content and Al content on the reverse transformation peak temperature for (A) Ti-18Zr-(12–14)Nb and (B) Ti-18Zr-(0–4)Al alloys.

1.3 CRYSTALLOGRAPHY OF MARTENSITIC TRANSFORMATION

1.3.1 Lattice Correspondence

For the cubic-orthorhombic martensitic transformation, there are six different lattice correspondences that have different orientation relationships with respect to the parent lattice. Fig. 1.13 shows six lattice correspondence variants (assigned as CV1, CV2, CV3, CV4, CV5, and CV6) of the α″ martensite. For example, the orientation relationships between the β and α″ phases for the CV1 variant are expressed as follows: $[100]_{\alpha''} - [100]_{\beta}$, $[010]_{\alpha''} - [011]_{\beta}$, and $[001]_{\alpha''} - [0\bar{1}1]_{\beta}$. The variant obtained by exchanging the *b*-axis and the *c*-axis of the α″ martensite variant CV1 is taken as CV2. The lattice correspondence for each variant is listed in Table 1.2. CV1 and CV2 have a twinning relationship that have a common axis of *a* and a twinning plane of $(011)_{\alpha''}$ or $(0\bar{1}1)_{\alpha''}$, i.e., CV1 and CV2 are in compound twinning relation [35]. CV3 and CV4 have a relationship in which the *a*-axis and the *b*-axis are mutually exchanged with a common axis of *a*. This crystal symmetry also results in the compound twining relationship. Similarly, CV5 and CV6 have the $\{011\}_{\alpha''}$ compound twinning relationship. On the other hand, CV1 has a twinning relationship by $\langle 211\rangle_{\alpha''}$ type II twinning or $\{111\}_{\alpha''}$ type I twinning with other variants except CV2. Similarly, the variants having no common axis of *a* are mutually related by $\langle 211\rangle_{\alpha''}$ type II twinning or $\{111\}_{\alpha''}$ type I twinning.

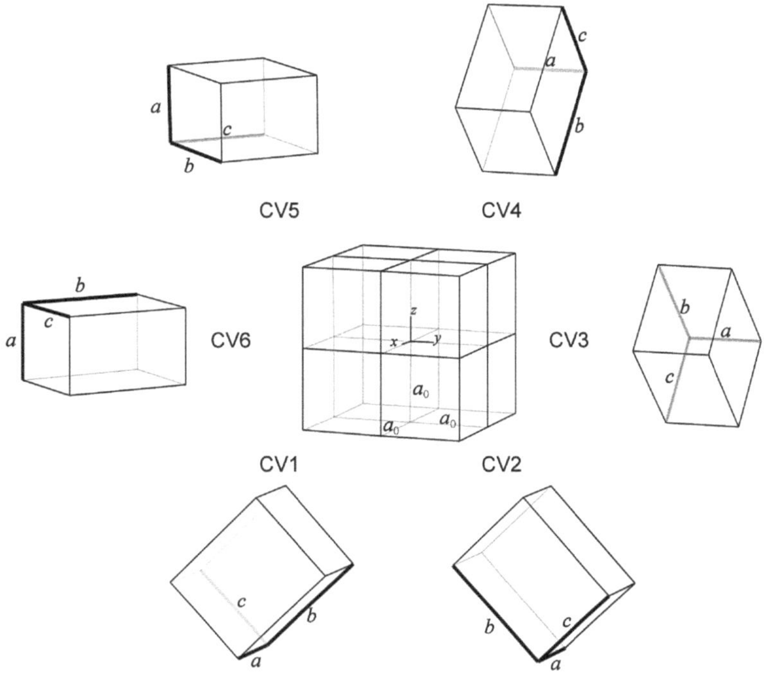

Fig. 1.13 Lattice correspondence variants of the α″ martensite and their orientation relationship.

Table 1.2 Lattice corresponding variants in β-α″ martensitic transformation

C.V.	$[100]_{\alpha''}$	$[010]_{\alpha''}$	$[001]_{\alpha''}$
CV1	$[100]_\beta$	$[011]_\beta$	$[0\bar{1}1]_\beta$
CV2	$[\bar{1}00]_\beta$	$[0\bar{1}1]_\beta$	$[011]_\beta$
CV3	$[010]_\beta$	$[101]_\beta$	$[10\bar{1}]_\beta$
CV4	$[0\bar{1}0]_\beta$	$[10\bar{1}]_\beta$	$[101]_\beta$
CV5	$[001]_\beta$	$[110]_\beta$	$[\bar{1}10]_\beta$
CV6	$[00\bar{1}]_\beta$	$[\bar{1}10]_\beta$	$[110]_\beta$

1.3.2 Transformation Strain

For the CV1 martensite variant, the lattice distortion matrix T_1' is expressed as follows in the coordinates of the α″ martensite phase using the lattice parameters of the β phase (a_0) and the α″ martensite phase (a, b, c):

$$T_1' = \begin{bmatrix} a/a_0 & 0 & 0 \\ 0 & b/\sqrt{2}a_0 & 0 \\ 0 & 0 & c/\sqrt{2}a_0 \end{bmatrix}. \tag{1.1}$$

The lattice distortion matrix of $\boldsymbol{T}_1$ in the coordinates of the β phase for the CV1 martensite variant is expressed as follows:

$$\boldsymbol{T}_1 = \boldsymbol{R}_1 \boldsymbol{T}'_1 \boldsymbol{R}_1^t \tag{1.2}$$

where $\boldsymbol{R}_1$ is a coordinate transformation matrix from the CV1 martensite variant coordinate frame to the β-phase coordinate frame and $\boldsymbol{R}_1^t$ is the transpose of $\boldsymbol{R}_1$, being a coordinate transformation matrix from the β-phase coordinate frame to the CV1 martensite variant coordinate frame. According to the lattice correspondence between the CV1 martensite variant and the β phase, $\boldsymbol{R}_1$ is expressed as follows:

$$\boldsymbol{R}_1 = \begin{bmatrix} 1 & 0 & 0 \\ 0 & 1/\sqrt{2} & -1/\sqrt{2} \\ 0 & 1/\sqrt{2} & 1/\sqrt{2} \end{bmatrix} \tag{1.3}$$

From Eqs. (1.1)–(1.3), the lattice distortion matrix of $\boldsymbol{T}_1$ is given as

$$\boldsymbol{T}_1 = \begin{bmatrix} a/a_0 & 0 & 0 \\ 0 & (b+c)/2\sqrt{2}a_0 & (b-c)/2\sqrt{2}a_0 \\ 0 & (b-c)/2\sqrt{2}a_0 & (b+c)/2\sqrt{2}a_0 \end{bmatrix} \tag{1.4}$$

Let a/a_0, $b/\sqrt{2}a_0$, and $c/\sqrt{2}a_0$ be α, β, and γ, respectively; $\boldsymbol{T}_1$ is displayed as

$$\boldsymbol{T}_1 = \begin{bmatrix} \alpha & 0 & 0 \\ 0 & \dfrac{\beta+\gamma}{2} & \dfrac{\beta-\gamma}{2} \\ 0 & \dfrac{\beta-\gamma}{2} & \dfrac{\beta+\gamma}{2} \end{bmatrix} \tag{1.5}$$

The coordinate transformation matrices and lattice distortion matrices for all six variant are listed in Tables 1.3.

Considering an arbitrary vector $\boldsymbol{x}$ in the coordinate of the β phase, it transforms to $\boldsymbol{x}'$ according to the following formula due to the martensitic transformation from the β phase to the CV1 martensite variant:

$$\boldsymbol{x}' = \boldsymbol{T}_1 \boldsymbol{x} \tag{1.6}$$

The transformation strain $\boldsymbol{\varepsilon}^i$ in an arbitrary direction can be calculated as follows:

$$\boldsymbol{\varepsilon}^i = (|\boldsymbol{x}'| - |\boldsymbol{x}|)/|\boldsymbol{x}| \tag{1.7}$$

For example, the transformation strain $\boldsymbol{\varepsilon}^i$ is calculated using the lattice parameters of Ti-20 at.% Nb; $a_0 = 0.3285\,\text{nm}$, $a = 0.3126\,\text{nm}$,

Table 1.3 Coordinate transformation matrices and lattice distortion matrices for six lattice corresponding variants

C.V.	Coordinate transformation matrix	Lattice distortion matrix
CV1	$R_1 = \begin{bmatrix} 1 & 0 & 0 \\ 0 & 1/\sqrt{2} & -1/\sqrt{2} \\ 0 & 1/\sqrt{2} & 1/\sqrt{2} \end{bmatrix}$	$T_1 = \begin{bmatrix} \alpha & 0 & 0 \\ 0 & \frac{\beta+\gamma}{2} & \frac{\beta-\gamma}{2} \\ 0 & \frac{\beta-\gamma}{2} & \frac{\beta+\gamma}{2} \end{bmatrix}$
CV2	$R_2 = \begin{bmatrix} -1 & 0 & 0 \\ 0 & -1/\sqrt{2} & 1/\sqrt{2} \\ 0 & 1/\sqrt{2} & 1/\sqrt{2} \end{bmatrix}$	$T_2 = \begin{bmatrix} \alpha & 0 & 0 \\ 0 & \frac{\beta+\gamma}{2} & \frac{-\beta+\gamma}{2} \\ 0 & \frac{-\beta+\gamma}{2} & \frac{\beta+\gamma}{2} \end{bmatrix}$
CV3	$R_3 = \begin{bmatrix} 0 & 1/\sqrt{2} & 1/\sqrt{2} \\ 1 & 0 & 0 \\ 0 & 1/\sqrt{2} & -1/\sqrt{2} \end{bmatrix}$	$T_3 = \begin{bmatrix} \frac{\beta+\gamma}{2} & 0 & \frac{\beta-\gamma}{2} \\ 0 & \alpha & 0 \\ \frac{\beta-\gamma}{2} & 0 & \frac{\beta+\gamma}{2} \end{bmatrix}$
CV4	$R_4 = \begin{bmatrix} 0 & 1/\sqrt{2} & 1/\sqrt{2} \\ -1 & 0 & 0 \\ 0 & -1/\sqrt{2} & 1/\sqrt{2} \end{bmatrix}$	$T_4 = \begin{bmatrix} \frac{\beta+\gamma}{2} & 0 & \frac{-\beta+\gamma}{2} \\ 0 & \alpha & 0 \\ \frac{-\beta+\gamma}{2} & 0 & \frac{\beta+\gamma}{2} \end{bmatrix}$
CV5	$R_5 = \begin{bmatrix} 0 & 1/\sqrt{2} & -1/\sqrt{2} \\ 0 & 1/\sqrt{2} & 1/\sqrt{2} \\ 1 & 0 & 0 \end{bmatrix}$	$T_5 = \begin{bmatrix} \frac{\beta+\gamma}{2} & \frac{\beta-\gamma}{2} & 0 \\ \frac{\beta-\gamma}{2} & \frac{\beta+\gamma}{2} & 0 \\ 0 & 0 & \alpha \end{bmatrix}$
CV6	$R_6 = \begin{bmatrix} 0 & -1/\sqrt{2} & 1/\sqrt{2} \\ 0 & 1/\sqrt{2} & 1/\sqrt{2} \\ -1 & 0 & 0 \end{bmatrix}$	$T_6 = \begin{bmatrix} \frac{\beta+\gamma}{2} & \frac{-\beta+\gamma}{2} & 0 \\ \frac{-\beta+\gamma}{2} & \frac{\beta+\gamma}{2} & 0 \\ 0 & 0 & \alpha \end{bmatrix}$

$b = 0.4870$ nm, and $c = 0.4646$ nm. The results for six variants along three representative orientations of [001], [011], and $[\bar{1}11]$ are listed in Table 1.4. The martensitic transformation results in a positive strain of 2.45% along the [001] direction for the CV1, CV2, CV3, and CV4. On the other hand, a negative strain of −4.84% is obtained for the CV5 and CV6, implying that the CV5 and CV6 are not favorable variants under the tensile stress but favorable under the compression. These results are in accordance with the lattice correspondence between the β phase and each martensite variant. For the CV5 and CV6, the [001] direction axis of the β phase is corresponding to the $[100]_{\alpha''}$ direction that shrinks due to the martensitic transformation. Only the CV1 exhibits a larger transformation strain along the [011] direction,

Table 1.4 Transformation strain for six variant along three representative orientations

	$\varepsilon^{[001]}$	$\varepsilon^{[011]}$	$\varepsilon^{[\bar{1}11]}$
CV1	0.0245	0.0483	0.0171
CV2	0.0245	0.0001	−0.0158
CV3	0.0245	−0.0113	−0.0158
CV4	0.0245	−0.0113	0.0171
CV5	−0.0484	−0.0113	−0.0158
CV6	−0.0484	−0.0113	0.0171

indicating that the CV1 is the most favorable variant to form when the tensile stress applied. Only the CV1 variant has a positive strain value along all orientations located in a $[001]-[011]-[\bar{1}11]$ standard stereographic triangle. Fig. 1.14 shows the orientation dependence of the transformation strain $\boldsymbol{\varepsilon}^i$ expressed by isostrain lines for the CV1. The transformation strain is the maximum along the [011] direction, and it decreases monotonically as the direction changes from [011] to [001] or $[\bar{1}11]$. It has been suggested that, if constraint between grains is ignored, the transformation strain of polycrystalline structure can be estimated as follows by averaging $\boldsymbol{\varepsilon}^i$ for representative 36 orientations that locate periodically in the standard stereographic triangle [78, 79]:

$$\bar{\varepsilon}=\left(\sum_{i=1}^{36}\varepsilon^i I^i\right)/36 \tag{1.8}$$

where I^i is orientation density that is depending on the loading direction.

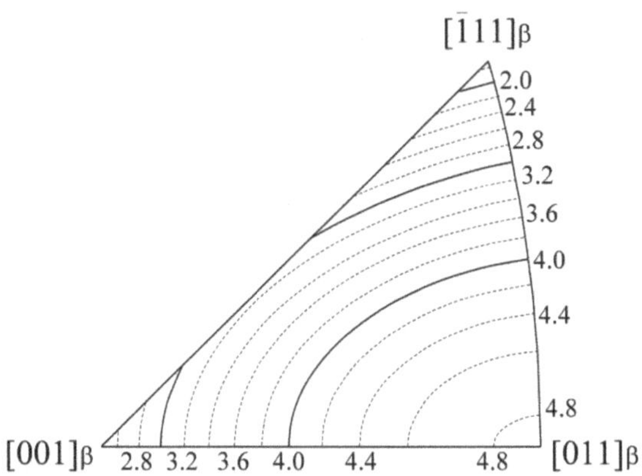

Fig. 1.14 Orientation dependence of the transformation strain for the martensite variant CV1.

1.4 MORPHOLOGY OF MARTENSITE PHASE

1.4.1 Habit Plane and Interface

Martensitic transformation proceeds on a specific interface (habit plane) between parent phase and martensite phase. Ideally, the habit plane should be undistorted and unrotated, i.e., invariant plane, to maintain coherency and to minimize the strain energy. Generally, a single variant of martensite does not satisfy the invariant condition; therefore, an additional lattice invariant process (lattice invariant shear, LIS) such as twinning and dislocation slip is necessary to form the invariant interface. It has been well established that the crystallographic characteristics of martensitic transformation including habit plane and invariant shear are well explained by the phenomenological theory of martensite crystallography (PTMC) [80–82,84]. According to PTMC, the matrix representing total shape change matrix or invariant shape strain matrix $\boldsymbol{P}_1$ is described by product of three matrices:

$$\boldsymbol{P}_1 = \boldsymbol{RPB} \tag{1.9}$$

where $\boldsymbol{B}$ is a lattice deformation matrix creating the martensite structure from the parent phase, $\boldsymbol{P}$ is a matrix representing lattice invariant shear in martensite, and $\boldsymbol{R}$ is a lattice rotation matrix of martensite structure. The shape strain $\boldsymbol{P}_1$ is also represented as

$$\boldsymbol{P}_1 = \boldsymbol{I} + \boldsymbol{m}_1 \boldsymbol{d}_1 \boldsymbol{p}_1' \tag{1.10}$$

where $\boldsymbol{I}$ is the 3×3 identity matrix, $\boldsymbol{m}_1$ is the magnitude of the total shape change, $\boldsymbol{d}_1$ is a unit column vector representing the direction of the total shape change, and $\boldsymbol{p}_1{}'$ is a unit row vector representing the direction normal to invariant plane, i.e., habit plane. The habit plane normal $\boldsymbol{p}_1{}'$ and other crystallographic parameters can be determined by solving Eq. (1.10) under the invariant plane strain condition on the habit plane. For the PTMC calculation, in addition to the lattice parameters of both the parent and martensite phases and their lattice correspondence, an appropriate lattice invariant shear is required.

Although the habit planes for the $\beta \rightarrow \alpha''$ martensitic transformation in β-Ti alloys have been investigated since the early 1950s, only a few quantitative studies have been performed. Table 1.5 lists some studies on the habit plane of Ti-Mo, Ti-Ta, and Ti-Nb alloys [4,6,31,35,36,84]. Experimental results showed that, although there were considerable scatter, the habit planes of α″ martensite in Ti-Mo alloys can be divided as two groups, i.e., {334} and {344}. Liu [4] reported that habit planes near {334} are

Table 1.5 Habit planes observed in Ti-Nb, Ti-Mo, and T-Ta-based alloys

Composition	Processing	Habit plane	
Ti-(10–11)Mo (wt%)	Ice-water quenching	~{334}, ~{344}	[4]
	Liquid argon quenching	~{344}	
Ti-10.5Mo (wt%)	Rolling	~{344}	
Ti-13Mo (wt%)	Compression	~{344}	
Ti-14Mo (wt%)	Tension	~{344}	
Ti-11Mo (wt%)	Compression	~{334}, ~{344}	[6]
Ti-12.5Mo (wt%)	Compression	~{344}	
Ti-53.2Ta (wt%) (Ti-21Ta (at.%))	Ice-water quenching	~{445}	[31]
Ti-20Nb (at.%)	Water quenching	~{557}	[36]
Ti-22Nb (at.%)	Water quenching	~{334}	
Ti-24Nb (at.%)	Water quenching	~{334}	
Ti-16Nb-3Al (at.%)	Water quenching	~{344}, ~{557}	[35]
Ti-18Nb-3Al (at.%)	Water quenching	~{111}	
Ti-23Nb-3Al (at.%)	Water quenching	~{557}	
Ti-27Nb (at.%)	Compression	~{0.47 0.51 0.72}	[84]

prevalently observed in water quenched Ti-(10 − 11)wt% Mo alloys. The {334} habit plane has been also reported to be dominant in water quenched Ti alloys such as Ti-Mn [2], Ti-Cr, and Ti-Fe [4]. On the other hand, the {344} habit planes were formed in Ti-10.5 wt% Mo and Ti-11 wt% Mo alloys after the subzero (87 K) quenching treatment. The {344} habit planes were also observed in Ti-(10.5–15)wt% Mo alloys deformed (tension, compression, and rolling) at room temperature. Liu [4] suggested that the habit plane is determined by the temperature where the martensite forms: the {334} habit plane is favorable in the relatively low alloyed Ti alloys that have a higher M_s and the stress-induced martensites form with the {344} habit planes. However, Gaunt and Christian [6] reported that both {334} and {344} habit planes were produced by stress-induced martensitic transformation in a Ti-11 wt% Mo alloy. They also mentioned that the {334} habit plane shows a good agreement with PTMC calculation.

Bywater and Christian [31] investigated the composition dependence of martensite morphologies in Ti-Ta alloys. They reported that the martensite in a Ti-10 at.% Ta alloy contains the $\{111\}_{\alpha''}$ type I internal twins as an LIS; on the other hand, untwinned (single variant) martensite is formed in a Ti-21 at.% Ta alloy. The difference of martensite morphology in the Ti-10 at.% Ta and Ti-21 at.% Ta alloys is well explained by the differences of the principal strains of the lattice deformation for the $\beta \rightarrow \alpha''$ martensitic transformation. The principal strains of the lattice deformation are given as

$$\eta_1 = (a - a_0)/a_0 \tag{1.11}$$

$$\eta_2 = \left(b - \sqrt{2}a_0\right)/\sqrt{2}a_0 \tag{1.12}$$

$$\eta_3 = \left(c - \sqrt{2}a_0\right)/\sqrt{2}a_0 \tag{1.13}$$

where a_0 is lattice parameter of the β phase and a, b, and c are the lattice parameters of the α″ martensite phase. In the Ti-21 at.% Ta alloy, one of the principal strains η_3 is close to zero, indicating that LIS is not required to form an invariant plane, i.e., habit plane. The observed habit plane in Ti-21 at.% Ta alloy is near {544} that is in good agreement with the predicted habit plane of {0.665 0.528 0.528} by PTMC without introducing LIS. However, composition dependence on the habit plane in Ti-Ta alloys has not been investigated in detail. Ivasishin et al. [85] reported a prediction that the habit plane in Ti-Mo and Ti-Ta alloys rotates from {434} to {433} with increasing Ta or Mo content based on minimization of elastic strain energy, but no experimental results were provided.

The morphologies of martensite in Ti-Nb-based alloys have been investigated by Chai et al. [36] and Inamura et al. [35]. Inamura et al. [35] systemically investigated the composition dependence of the morphology of martensite and habit plane normal in Ti-(10 − 30)Nb-3Al and Ti-24Nb alloys. They showed that the magnitude of η_3 is a determining parameter for the morphology of martensite and habit plane normal. Fig. 1.15 shows the Nb content dependence of the principal strains of lattice deformation for the Ti-xNb and Ti-xNb-3Al alloys [16,35]. The signs of η_1 and η_2 are opposite, but the magnitudes are similar. The magnitudes of both η_1 and η_2 decrease linearly with increasing Nb content. The magnitude of η_3 is small when compared with other principal strains. It is noted that η_3 decreases with increasing Nb content and becomes zero at 20 at.% Nb for Ti-Nb binary alloys and at 23 at.% Nb for Ti-Nb-3Al ternary alloys, and then, its sign becomes negative with further increasing Nb content. According to PTMC analysis, in order to form an invariant plane, $\{111\}_{\alpha''}$ type I twinning or $\langle 211\rangle_{\alpha''}$ type II twinning can be LIS when η_3 is positive, while $\{011\}_{\alpha''}$ compound twinning can be LIS when η_3 is negative. On the other hand, LIS is not necessary when η_3 is zero.

In the Ti-16Nb-3Al alloy ($\eta_3 = 0.0077$), two types of martensite morphology were observed. One is an internal twinned martensite, and the other is a nontwinned single variant martensite. Fig. 1.16 shows a typical bright-field image of twinned martensites and the corresponding selected area diffraction

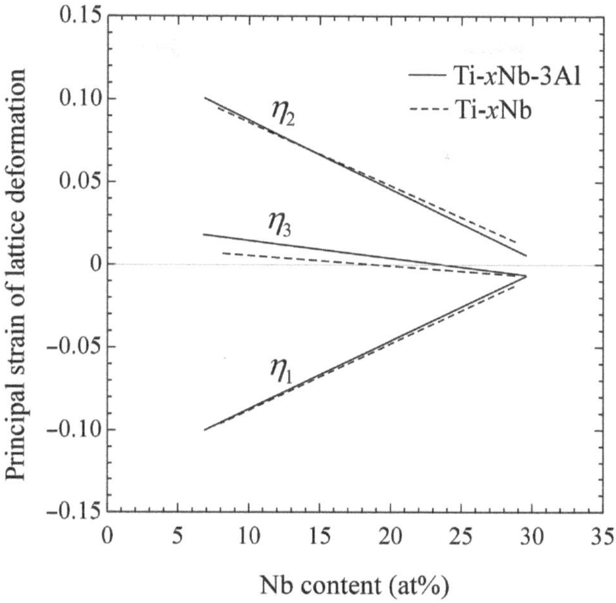

Fig. 1.15 Nb content dependence of the principal strains of lattice deformation for the Ti-*x*Nb-3Al alloys.

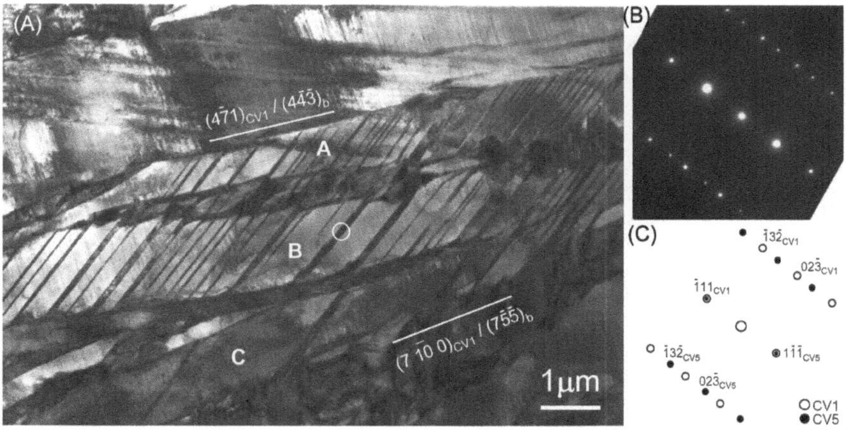

Fig. 1.16 (A) A bright-field image of twinned martensites and (B) the corresponding selected area diffraction pattern obtained in a Ti-16Nb-3Al alloy. (C) Key diagram of (B).

pattern (SADP) [35]. The SADP taken in the encircled region exhibits that the internal twin is $\{111\}_{\alpha''}$ type I where CV1 is the major variant and CV5 is a twinning variant. The average volume fraction of internal twins is measured to be 0.27, 0.24, and 0.08 for plate A, B, and C, respectively. It is noted that the

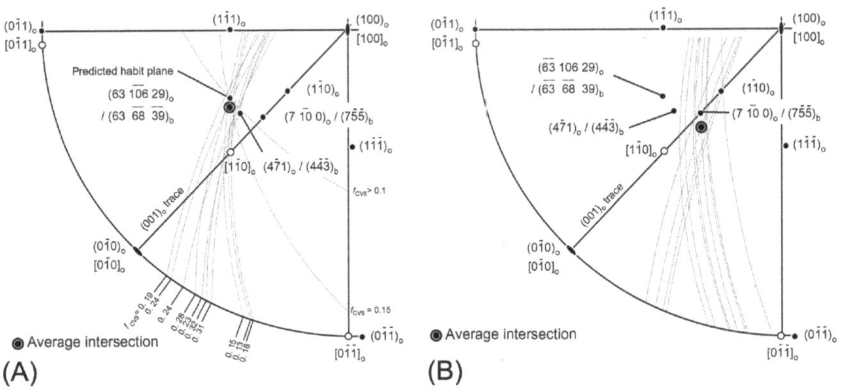

Fig. 1.17 Single trace analysis for the determination of the habit plane for the (A) twinned martensites and (B) nontwinned martensites in a Ti-16Nb-3Al alloy.

habit plane trace is different for plate A and C. The habit plane trace of A is close to $\left(4\overline{43}\right)$, but that of C is close to $\left(7\overline{55}\right)$. Fig. 1.17A shows the results of single trace analysis for the determination of the habit plane of twinned martensites [35]. The average volume fraction of internal twins in the twinned martensites is 0.22. The average habit plane normal of twinned martensites is determined to be $\left(0.61\,\overline{0.67}\,\overline{0.43}\right)$, which is 4° away from $\left(4\overline{43}\right)$. As shown in Fig. 1.17A, the average habit plane normal is close to that predicted from PTMC. On the other hand, Fig. 1.17B shows that the average habit plane normal of nontwinned martensites locates around $\left(7\overline{55}\right)$ that is relatively far from the predicted habit plane by PTMC. These results are reasonable, since LIS is unavoidable to satisfy the invariant plane condition of the habit plane.

Fig. 1.18A shows a typical BFI of martensite structure of the partially transformed Ti-23Nb-3Al alloy ($\eta_3 = 0$) [35]. It is clearly seen that each martensite plate is consisted of a single variant with no internal twin. As shown in Fig. 1.18D, the habit planes determined from edge-on directions of martensite plates are located near $\left(7\overline{55}\right)$. The average habit plane normal is determined to be $\left(0.75\,\overline{0.47}\,\overline{0.47}\right)$. These results are well consistent with PTMC analysis. The PTMC analysis of Ti-23Nb-3Al gives a habit plane solution of $\left(0.7051\,\overline{0.5014}\,\overline{0.5014}\right)$ without any LIS (internal twinning). The nontwinned martensites are mainly observed in the Ti-18Nb-3Al and Ti-24Nb alloys where η_3 is not zero ($\eta_3 = 0.0055$ in Ti-18Nb-3Al, and $\eta_3 = -0.0018$ in Ti-24Nb). In these alloys, the habit plane is close to $\left(1\overline{11}\right) - \left(7\overline{55}\right)$ that is slightly different from the predicted one by PTMC.

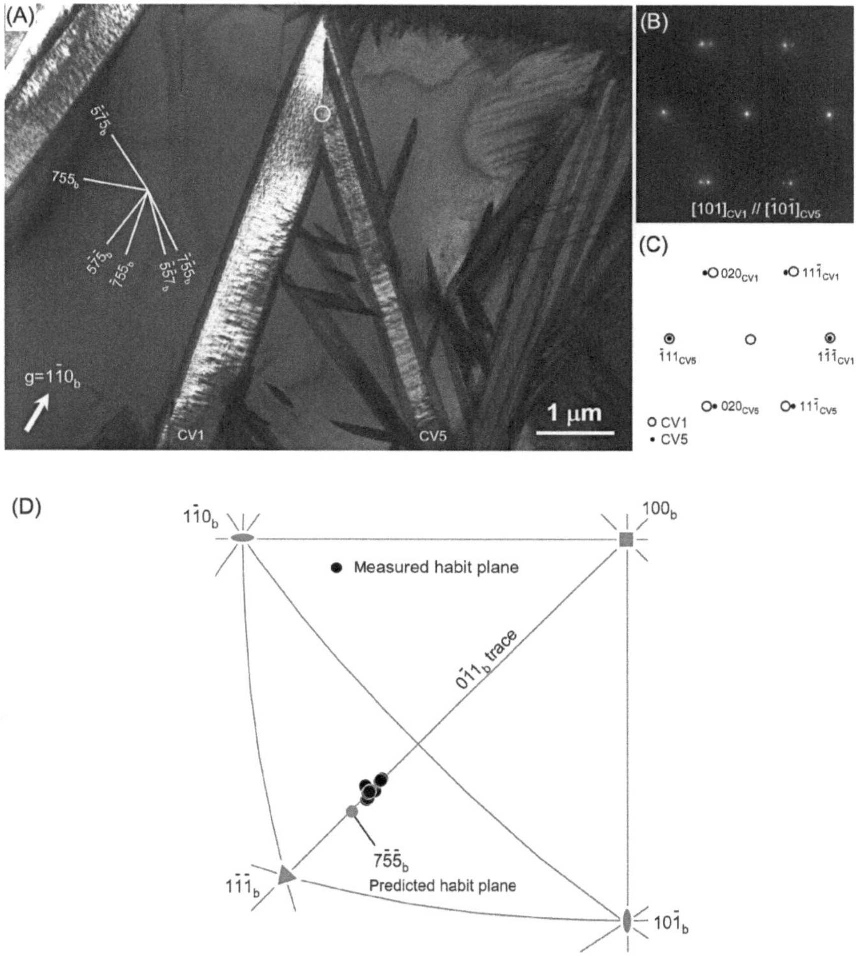

Fig. 1.18 (A) A bright-field image of martensite structure of the partially transformed Ti-23Nb-3Al alloy. (B) Selected area diffraction pattern taken from the interface area (encircled region of (A)). (C) Key diagram of (B). (D) Habit plane analysis.

Inamura et al. [35] claimed that a critical value of the magnitude η_3, i.e., a critical value of elastic strain energy at the habit plane, is required for the formation of the internal twin as an LIS, because the formation of internal twin is accompanied by additional energy of twin planes.

Tahara et al. [84] reported the habit planes of the stress-induced martensite plates in a Ti-27Nb single crystal. They showed that a single variant of martensite plate can be obtained by compression if the compression axis is properly selected. For example, plate-shaped α'' plates with a single variant of V5 are formed when the compression axis is $[\bar{1}48]$, where the habit plane is closed to (0.47 0.51 0.72). The observed habit planes are confirmed to be in good agreement with the predicted one from PTMC. They also noted that no internal twins are formed in the stress-induced martensite plates.

Even though the PTMC well describes the morphology of martensitic transformation, it does not include the atomic-scale mechanism. Chai et al. [36] investigated the structure of the β/α'' interfaces in Ti-Nb binary alloys. Fig. 1.19 shows bright-field images and selected diffraction patterns of Ti-20Nb, Ti-22Nb, and Ti-24Nb alloys where the α'' martensite plates are viewed perfectly edge-on [36]. In the Ti-20Nb alloy, the edge-on α'' martensite plate (α''_{V2}) is observed parallel to $[001]_{\alpha''}$. It is seen that the β phase remained in the Ti-22Nb and Ti-24Nb alloys; the edge-on α'' martensite

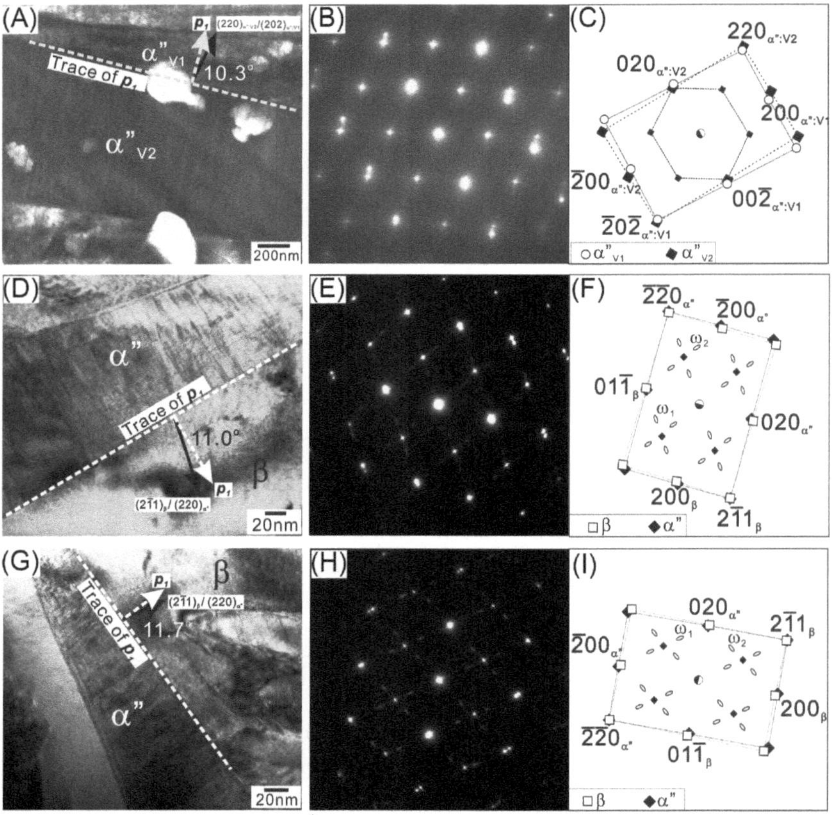

Fig. 1.19 Bright-field images ((A), (D), and (G)) and selected diffraction patterns ((B), (E), and ((H)) of Ti-20Nb, Ti-22Nb, and Ti-24Nb alloys, respectivley. (C), (F), and (I) are key diagrams of (B), (E), and (H), respectively.

plate and the β phase are observed along $[001]_{\alpha''}$ and $[0\bar{1}\bar{1}]_{\beta}$, respectively. From the selected area diffraction patterns, it is noticed that the interface plane normal is slightly dependent on the composition: 10.3° away from $(110)_{\alpha''}$ in the Ti-20Nb alloy, 11.0° away from $(110)_{\alpha''}/(2\bar{1}1)_{\beta}$ in the Ti-22Nb alloy, and 11.7° away from $(110)_{\alpha''}/(2\bar{1}1)_{\beta}$ in the Ti-24Nb alloy.

Fig. 1.20 shows high-resolution images of the Ti-20Nb, Ti-22Nb, and Ti-24Nb alloys [36]. It is seen that not only α''/α'' interface but also β/α'' interface is formed by a series of terrace and step at atomic scale that is consistent with the proposition by the topological model (TM) [86]. Fig. 1.20A shows that the terrace planes of the α''/α'' interface are parallel to $(101)_{\alpha''}$ and $(110)_{\alpha''}$ for the α''_{V1} and α''_{V2}, respectively. The α''/α'' interface is semicoherent, and a strong strain contrast at the interface is supposed to be due to

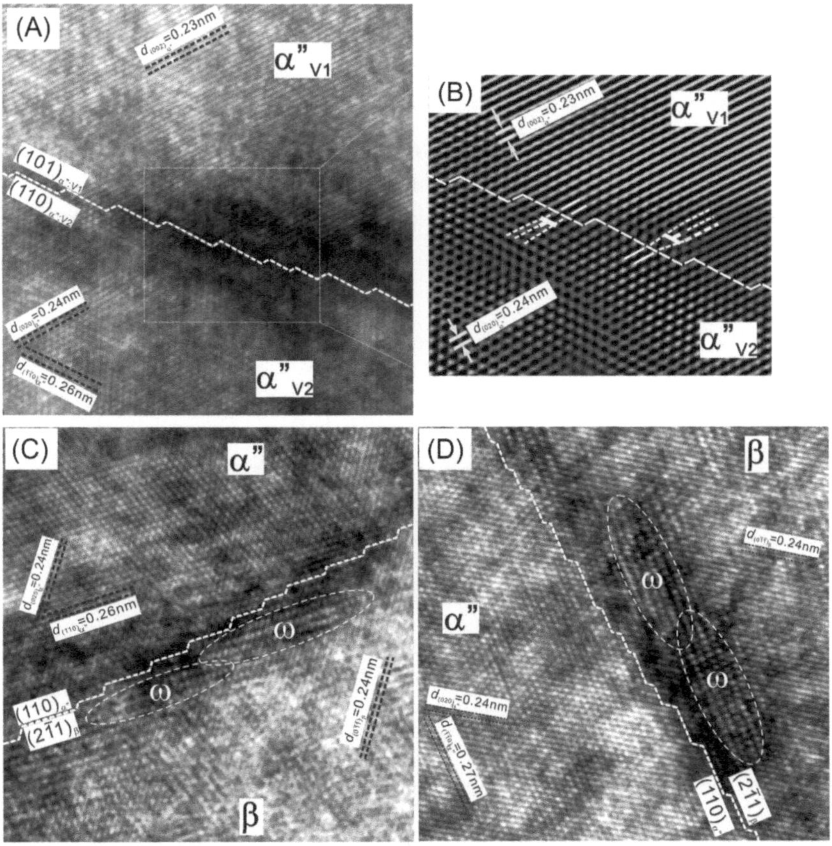

Fig. 1.20 High-resolution images of the (A) Ti-20Nb, (C) Ti-22Nb, and (D) Ti-24Nb alloys. (B) is an enlarged image of the white square area in (A).

the misfit strain resulting from different plane spacing. In the β/α'' interface, the terrace planes are parallel to $(2\bar{1}1)_{\beta}/(110)_{\alpha''}$, and the average lengths of the terraces are about 1.5 and 1.3 nm for the Ti-22Nb and Ti-24Nb alloys. The steps are consisted of one or two $(110)_{\alpha''}$ plane spacing. Chai et al. [36] have claimed that transformation disconnections (***b***, h) are superimposed on terrace interfaces to accommodate misfit strain, and the martensitic transformation proceeds by the lateral motion of an array of the interfacial defects. The magnitude of ***b***, h and the spacing of the transformation disconnections are found to be dependent on the Nb content. The theoretical prediction based on TM shows that the habit plane in Ti-20Nb alloy is close to $(7\bar{5}\bar{5})$ and it changes to $(4\bar{3}\bar{3})$ as the Nb content increases from 20 to 24 at.%. These results agreed well with those predicted from the phenomenological theory and also from the experimental observation as shown in Fig. 1.21 [36].

1.4.2 Self-Accommodation

According to the PTMC analysis, there are 12 solutions of habit plane variants (HPVs) in the $\beta \rightarrow \alpha''$ martensitic transformation in β-type Ti-based alloys. For example, 12 habit plane solutions for the Ti-20Nb alloy, designated as 1(+), 1(−), 2(+), 2(−), 3(+), 3(−), 4(+), 4(−), 5(+), 5(−), 6(+), and

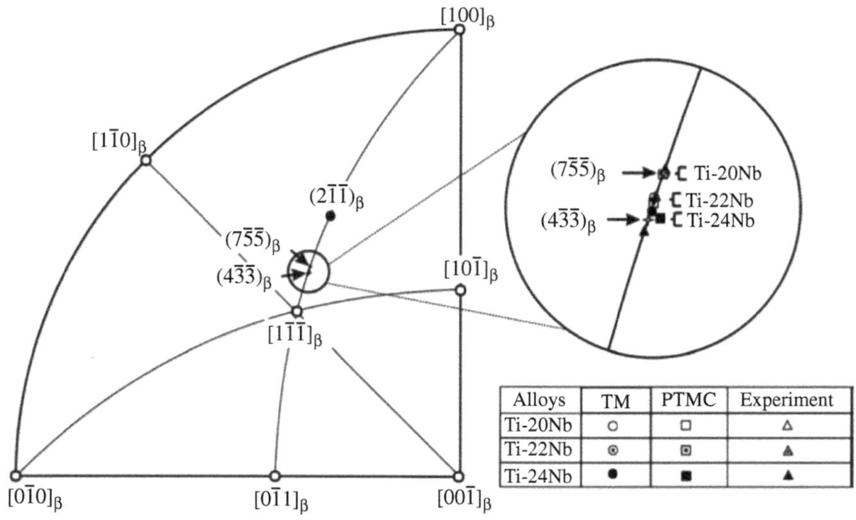

Fig. 1.21 Habit planes Ti-20Nb, Ti-22Nb, and Ti-24Nb predicted from the topological mode (TM) and phenomenological theory of martensite crystallography (PTMC) and experimental observation.

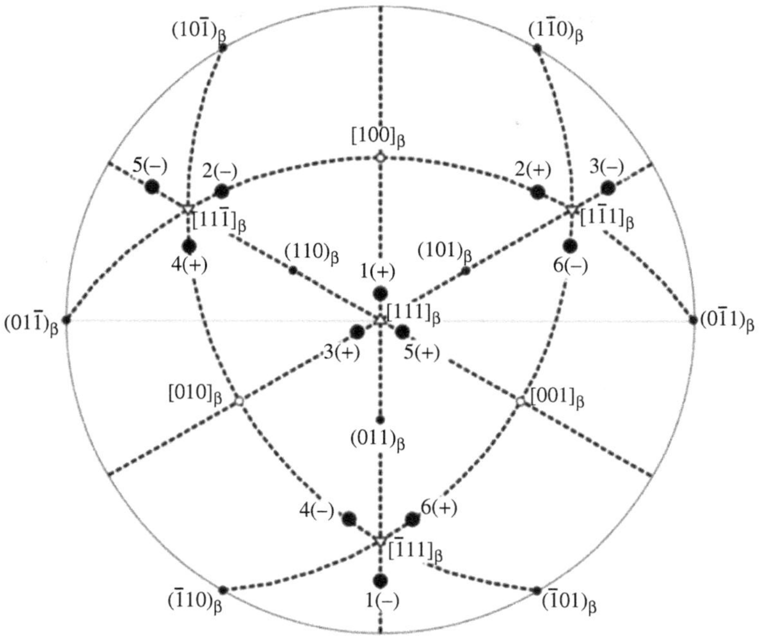

Fig. 1.22 Habit planes calculated form PTMC analysis for the Ti-20Nb alloy.

6(−), are shown in Fig. 1.22 [37]. It is seen that three habit plane variants are placed around each of ⟨111⟩ poles. The habit plane and the corresponding total shape change matrix $\boldsymbol{P}_1$ predicted by PTMC analysis [37] are listed in Table 1.6.

As a common characteristic of the martensitic transformation, self-accommodation morphologies have been reported in the α″ martensites. Fig. 1.23 shows a typical example of self-accommodation morphology consisting of three α″ martensite plates in the Ti-20Nb alloy [37]. A hollow triangular morphology is composed by the habit plane variants 1(−), 3(−), and 5(−). A solid triangular morphology on the upper right part is also composed by the habit plane variants 1(−), 3(−), and 5(−). Similar morphologies composed by three habit plane variants have been reported in Ti-Ni alloys [87]. Fig. 1.23B shows that two adjacent α″ plates are related by twinning on {111} and the habit planes are closed to {755}, which is consistent with the PTMC analysis. A V-shaped morphology is another common self-accommodation morphology. Fig. 1.24 shows an example of V-shaped morphology composed of two variants of α″ martensite plates, 4(+) and 6 (−) [37].

Table 1.6 Habit planes and the corresponding total shape change matrix predicted by PTMC analysis for the Ti-20Nb alloy

Habit plane variant	Habit plane	Invariant plane strain
1(+)	(0.6995 0.5057 0.5049)	$\begin{pmatrix} 0.9503 & -0.0359 & -0.0358 \\ 0.0326 & 1.0236 & 0.0235 \\ 0.0326 & 0.0236 & 1.0235 \end{pmatrix}$
1(−)	$(0.6992\ \overline{0.5055}\ \overline{0.5056})$	$\begin{pmatrix} 0.9503 & 0.0359 & 0.0359 \\ -0.0326 & 1.0236 & 0.0236 \\ -0.0325 & 0.0235 & 1.0235 \end{pmatrix}$
2(+)	$(0.6995\ \overline{0.5049}\ 0.5057)$	$\begin{pmatrix} 0.9503 & 0.0359 & -0.0359 \\ -0.0326 & 1.0235 & -0.0236 \\ 0.0326 & -0.0235 & 1.0236 \end{pmatrix}$
2(−)	$(0.6992\ 0.5056\ \overline{0.5054})$	$\begin{pmatrix} 0.9503 & -0.0359 & 0.0359 \\ 0.0325 & 1.0235 & -0.0235 \\ -0.0326 & -0.0236 & 1.0236 \end{pmatrix}$
3(+)	(0.5049 0.6995 0.5057)	$\begin{pmatrix} 1.0235 & 0.0326 & 0.0236 \\ -0.0359 & 0.9503 & -0.0359 \\ 0.0235 & 0.0326 & 1.0236 \end{pmatrix}$
3(−)	$(\overline{0.5056}\ 0.6992\ \overline{0.5054})$	$\begin{pmatrix} 1.0235 & -0.0325 & 0.0235 \\ 0.0359 & 0.9503 & 0.0359 \\ 0.0236 & -0.0326 & 1.0236 \end{pmatrix}$
4(+)	$(0.5057\ 0.6995\ \overline{0.5049})$	$\begin{pmatrix} 1.0236 & 0.0326 & -0.0235 \\ -0.0359 & 0.9503 & 0.0359 \\ -0.0236 & -0.0326 & 1.0235 \end{pmatrix}$
4(−)	$(\overline{0.5054}\ 0.6992\ 0.5056)$	$\begin{pmatrix} 1.0236 & -0.0326 & -0.0236 \\ 0.0359 & 0.9503 & -0.0359 \\ -0.0235 & -0.0325 & 1.0235 \end{pmatrix}$
5(+)	(0.5057 0.5049 0.6995)	$\begin{pmatrix} 1.0236 & 0.0235 & 0.0326 \\ 0.0236 & 1.0235 & 0.0326 \\ -0.0359 & -0.0359 & 0.9503 \end{pmatrix}$
5(−)	$(\overline{0.5054}\ \overline{0.5056}\ 0.6992)$	$\begin{pmatrix} 1.0236 & 0.0236 & -0.0326 \\ 0.0235 & 1.0235 & -0.0325 \\ 0.0359 & 0.0359 & 0.9503 \end{pmatrix}$
6(+)	$(\overline{0.5049}\ 0.5057\ 0.6995)$	$\begin{pmatrix} 1.0235 & -0.0236 & -0.0326 \\ -0.0235 & 1.0236 & 0.0326 \\ 0.0359 & -0.0359 & 0.9503 \end{pmatrix}$
6(−)	$(0.5056\ \overline{0.5054}\ 0.6992)$	$\begin{pmatrix} 1.0235 & -0.0235 & 0.0325 \\ -0.0236 & 1.0236 & -0.0326 \\ -0.0359 & 0.0359 & 0.9503 \end{pmatrix}$

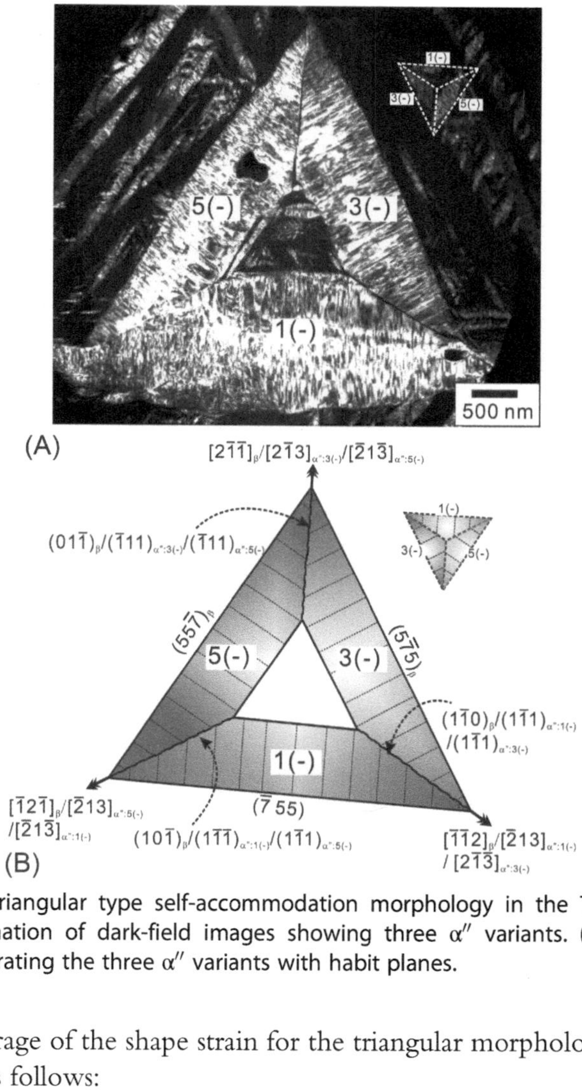

Fig. 1.23 A triangular type self-accommodation morphology in the Ti-20Nb alloy. (A) A combination of dark-field images showing three α'' variants. (B) Schematic diagram illustrating the three α'' variants with habit planes.

The average of the shape strain for the triangular morphology $\overline{\boldsymbol{P}}_1(\Delta)$ is calculated as follows:

$$\overline{\boldsymbol{P}}_1(\Delta) = \frac{\{\boldsymbol{P}_1(1(-)) + \boldsymbol{P}_1(3(-)) + \boldsymbol{P}_1(5(-))\}}{3}$$

$$= \begin{pmatrix} 0.9991 & 0.0090 & 0.0089 \\ 0.0089 & 0.9991 & 0.0090 \\ 0.0090 & 0.0089 & 0.9991 \end{pmatrix} \qquad (1.14)$$

where $\boldsymbol{P}_1(1(-))$, $\boldsymbol{P}_1(3(-))$, and $\boldsymbol{P}_1(3(-))$ are the total shape strain for the habit plane variants 1(−), 3(−), and 5(−), respectively. When comparing

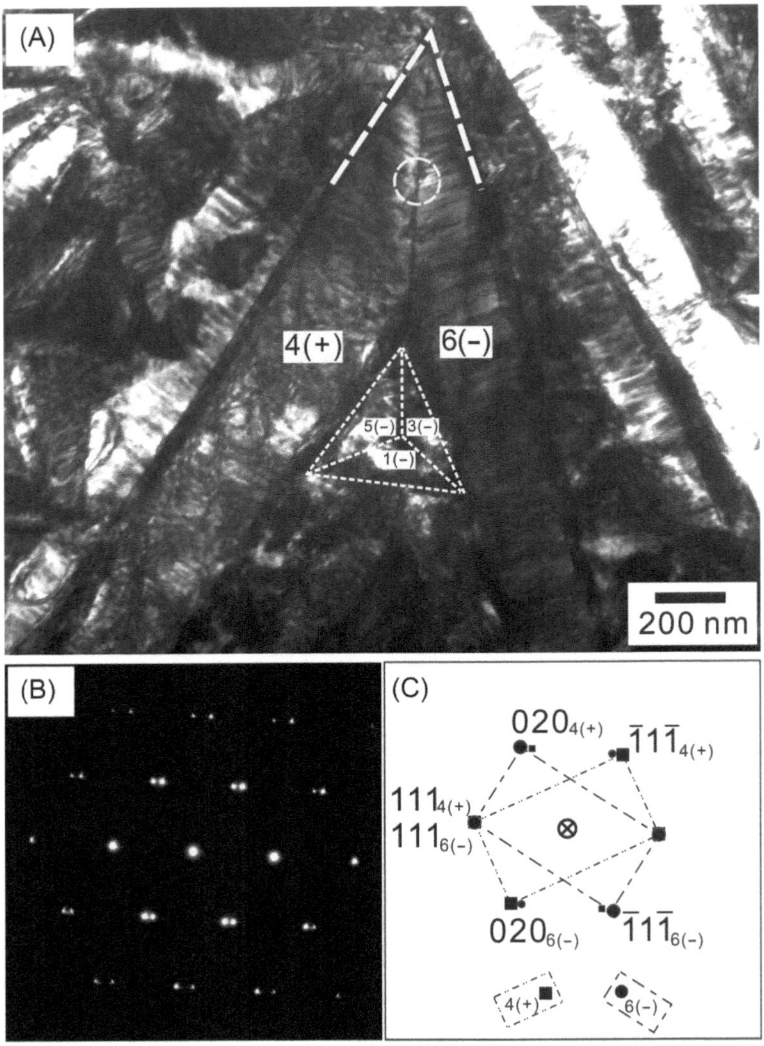

Fig. 1.24 (A) A bright-field image showing V-shaped self-accommodation morphology in the Ti-20Nb alloy. (B) Selected area diffraction pattern taken from the interface area. (C) Key diagram of (B).

$\overline{\boldsymbol{P}}_1(\Delta)$ with $\boldsymbol{P}_1(1(-))$, $\boldsymbol{P}_1(3(-))$, and $\boldsymbol{P}_1(3(-))$ in Table 1.5, it is very evident that the diagonal elements are closer to unity and the off-diagonal elements are closer to zero. This result indicates that not only the shear strain components but also the dilatation component decreases remarkably by clustering three variants, and the formation of the triangular morphology is favorable for the self-accommodation. Similarly, the average of the shape strain for the V-shaped morphology $\overline{\boldsymbol{P}}_1(\mathrm{V})$ is calculated as follows:

$$\overline{\boldsymbol{P}}_1(\mathrm{V}) = \frac{\{\boldsymbol{P}_1(4(+)) + \boldsymbol{P}_1(6(-))\}}{2} = \begin{pmatrix} 1.0236 & 0.0045 & 0.0045 \\ -0.0297 & 0.9870 & 0.0017 \\ -0.0297 & 0.0017 & 0.9869 \end{pmatrix} \quad (1.15)$$

where $\boldsymbol{P}_1(4(+))$ and $\boldsymbol{P}_1(6(-))$ are the total shape strain for the habit plane variants 4(+) and 6(−), respectively. The average shape strain decreases by the formation of the V-shaped morphology, but the change is not significant when comparing with $\boldsymbol{P}_1$ for a single variant, indicating that the V-shaped morphology is relatively unfavorable in comparison with the triangular morphology. It should be noticed that a small solid triangular morphology, consisting of three martensite variants 1(−), 3(−), and 5(−), exists between the two α'' martensite plates. Considering the contribution of the triangular morphology on the shape strain, the average of the shape strain $\overline{\boldsymbol{P}}_1(\mathrm{V}+\Delta)$ is calculated as follows:

$$\overline{\boldsymbol{P}}_1(\mathrm{V}+\Delta) = \frac{\{\boldsymbol{P}_1(4(+)) + \boldsymbol{P}_1(6(-)) + \boldsymbol{P}_1(1(-)) + \boldsymbol{P}_1(3(-)) + \boldsymbol{P}_1(5(-))\}}{5} = \begin{pmatrix} 1.0089 & 0.0072 & 0.0072 \\ -0.0065 & 0.9943 & 0.0061 \\ -0.0065 & 0.0060 & 0.9942 \end{pmatrix}. \quad (1.16)$$

It is clear that both shear and dilation components are reduced considerably by the combination of the V-shaped morphology and the triangular morphology consisting of five habit plane variants. As a result, it is suggested that the triangular morphology is introduced to reduce the total shape strain after the formation of the V-shaped morphology.

1.4.3 Antiphase Boundary-Like Stacking Fault

Fig. 1.25 shows bright-field images of α'' martensites observed along different zone axes, i.e., $[\bar{1}10]$, [110], and [001], in a Ti-23Nb-3Al alloy [38]. In all images, antiphase boundary (APB)-like faults are observed. Particularly, the APB-like domain contrast is clearly seen in Fig. 1.25B taken along the $[\bar{1}10]$ zone axis. Fig. 1.25C and D show a ribbonlike wavy contrast lying perpendicular to the habit plane. Similar APB-like fault morphologies have been reported in other Ti alloys such as Ti-Mo [21], Ti-Mn [88], Ti-V- Al [33],

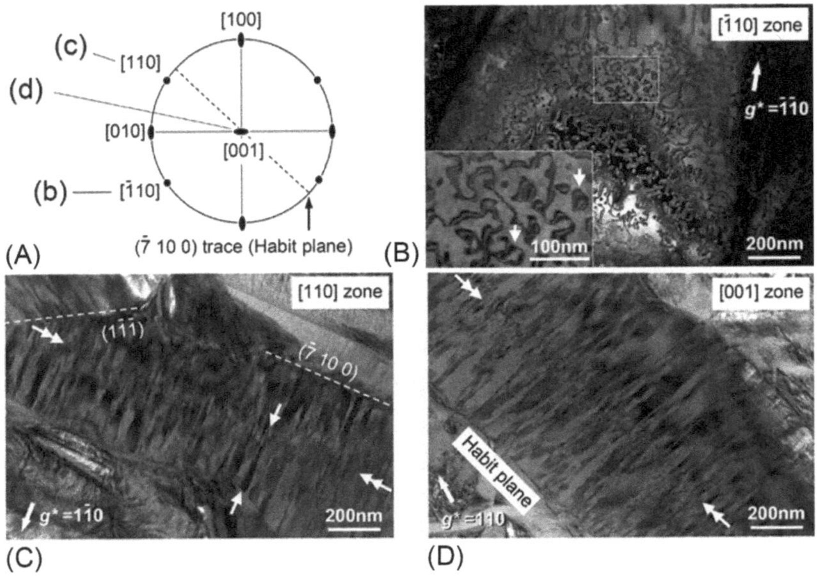

Fig. 1.25 Bright-field images of α″ martensites observed along different zone axes in a Ti-23Nb-3Al alloy. (A) 001 stereographic projection of standard variant of α″ martensite and bright-field images of α″ martensites observed near (B) $[\bar{1}10]$ zone, (C) [110] zone, and (D) [001] zone.

Ti-Nb [22], and Ti-V-Fe-Al [34]. Pak [33] assumed that the martensite phase of a Ti-15.4V-4Al (wt%) alloy has an ordered face-centered orthorhombic structure (*Cmmm*) on the basis of the appearance of APB-like faults and forbidden diffraction spots. On the other hand, Ahmed and Rack [22] proposed that ordering of orthorhombic α″ martensite occurs after the martensitic transformation β → α″ in Ti-Nb alloys. They also suggested that the ordered structure is a base-centered orthorhombic structure (*Pmmm*) as evidenced by the presence of superlattice reflections at 00l (l = odd numbers) in SADPs. However, it has been well confirmed that the α″ martensite is disordered as mentioned above. It is supposed that the forbidden superlattice reflections are due to the double diffraction and APB-like faults are not due to the ordering of α″ phase.

The APB-like morphology can be explained by the displacement induced fault due to two different modes of shuffling in the opposite directions during the β → α″ martensitic transformation [34, 38]. During the β → α″ martensitic transformation, shuffling occurs in such a way that the atoms in the (002) plane in the orthorhombic unit cell are shifted along the b-axis as shown in Fig. 1.26 [34]. There are two possible shuffling modes whose

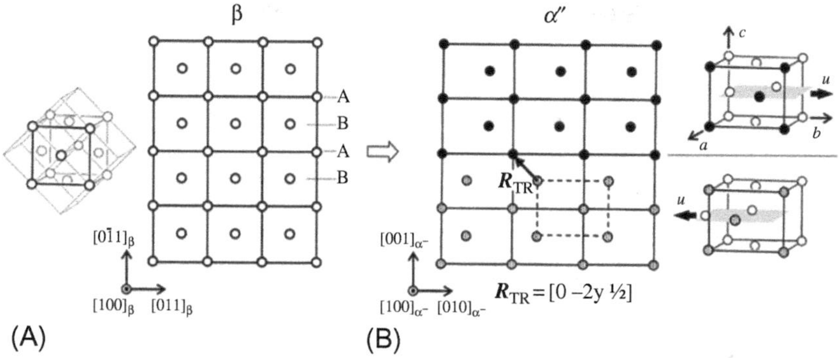

Fig. 1.26 Schematic explanation of the formation of the APB-like morphology. (A) $(100)_\beta$ layer of bcc structure and (B) (100) layer of α″ martensite.

directions are opposite to each other as shown in Fig. 1.26B: the shuffling direction is $[010]_{\alpha''}$ in upper half part and $[0\bar{1}0]_{\alpha''}$ in lower half part. It is supposed that two shuffling modes nucleate in a random manner since they are crystallographically and energetically equivalent. The boundary of these two shuffling modes results in the staking faults similar to APB as shown in Fig. 1.25B. A similar mechanism, i.e., transformation-induced APB or the displacement induced fault, is found to be operated in the B19 martensite of a Ti-Pd alloy [89].

1.5 OMEGA PHASE TRANSFORMATION

1.5.1 Crystal Structure and Lattice Correspondence

There has been considerable interest on the omega phase transformation (β → ω) and the nature of the ω phase, and there have been many excellent reviews on this topic [39,90–93]. This book focuses specifically on the effect of the omega phase on the martensitic transformation and the shape memory properties of Ti alloys. The omega phase transformation can be classified into two different types. One is martensitic, i.e., diffusionless and displacive, and therefore cannot be suppressed by rapid quenching. The other is thermally activated or diffusion controlled transformation. The ω phase formed by the former transformation is generally referred to as athermal ω (ω_{ath}) phase and has the same composition as the β phase. On the other hand, the thermally activated transformation involves a change in composition, and the product is denoted by aged or thermal ω (ω_{th}) phase. The crystal structures of both athermal and thermal ω phases are essentially identical although the compositions are different.

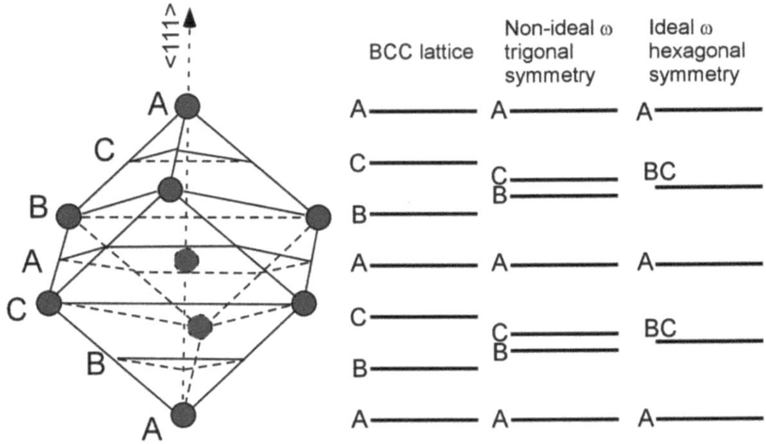

Fig. 1.27 Schematic explanation of {111} plane collapsing for the β → ω transformation.

The ω phase transformation is expressed by the collapsing of {222} planes in the bcc structure of the β phase. Along the ⟨111⟩ direction of the bcc structure, {222} planes are stacked as in a sequence of ABCABC as shown in Fig. 1.27. On the ideal β → ω transformation, the B and C planes collapse to one plane BC, resulting in hexagonal symmetry with the space group of *P6/mmm*. The collapse of the {222} planes can be considered as a ⟨111⟩ longitudinal displacive wave. When the collapse of the B and C planes is not complete, the ω phase has a trigonal structure ($P\overline{3}m1$). As shown in Fig. 1.28 [38], the β → ω transformation can be

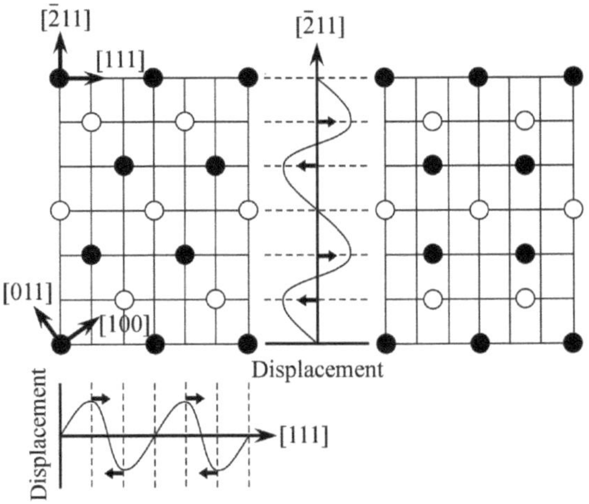

Fig. 1.28 Schematic explanation of a transverse modulation wave of $\{\overline{2}11\}\langle111\rangle$ for the β → ω transformation.

regarded as a transverse modulation wave of $\{\bar{2}11\}\langle 111\rangle$ that is crystallographically equivalent to the collapse of the $\{222\}$ planes. In this case, the wavelength is $a_\beta(3/2)^{1/2}$, and the amplitude of the modulation is $\sqrt{3}/12$ for the ideal ω phase with a hexagonal structure. And the following relationship is obtained: $d_{(0001)\omega} = 3d_{(222)\beta} = a_\beta\sqrt{3}/2$.

For the $\beta \rightarrow \omega$ transformation, there are four lattice correspondences, since the bcc structure has four $\langle 111\rangle$ directions, as shown in Fig. 1.29. The orientation relationships in four variants are listed in Table 1.7. For example, the orientation relationships between the β and ω phases for the

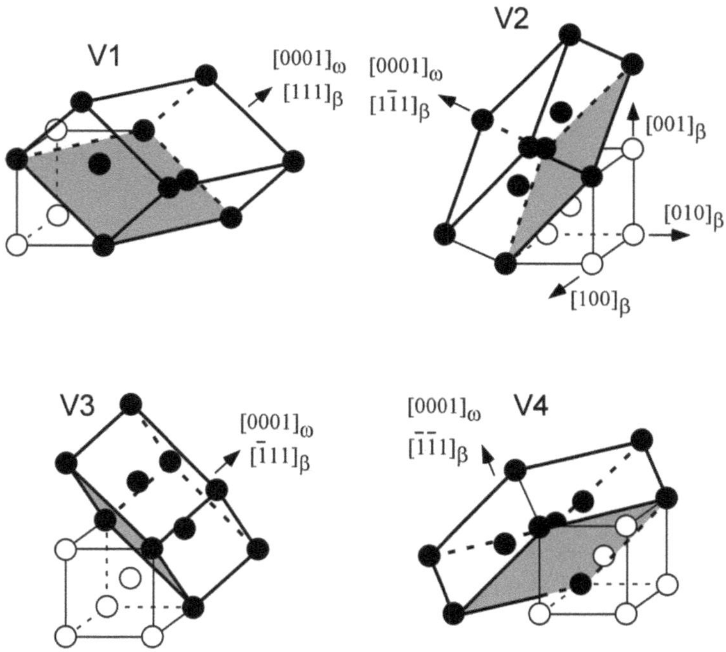

Fig. 1.29 Lattice correspondence variants of the ω phase and their orientation relationship.

Table 1.7 Lattice corresponding variants in β-ω transformation

Variant	$(0001)_\omega$	$\langle 11\bar{2}0\rangle_\omega$
V1	$(111)_\beta$	$[1\bar{1}0]_\beta$
V2	$(1\bar{1}1)_\beta$	$[\bar{1}01]_\beta$
V3	$(\bar{1}11)_\beta$	$[01\bar{1}]_\beta$
V4	$(\bar{1}\bar{1}1)_\beta$	$[\bar{1}10]_\beta$

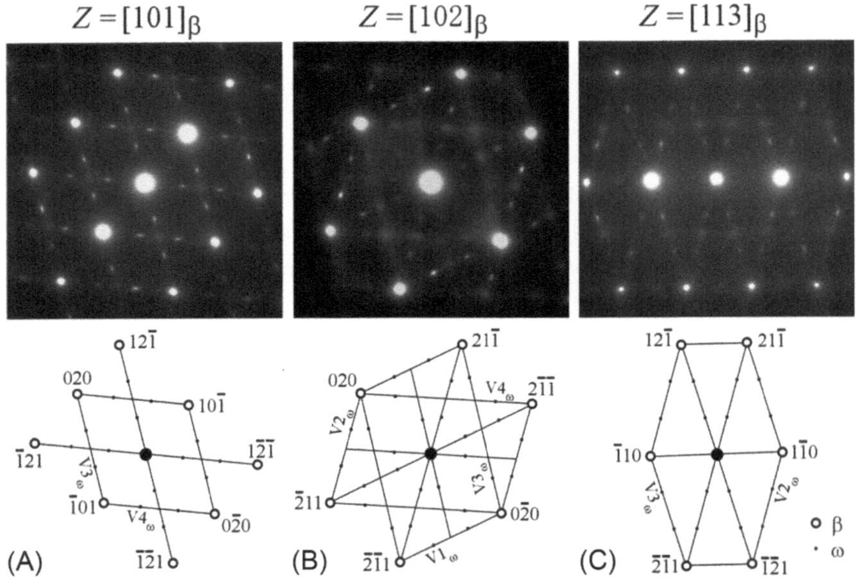

Fig. 1.30 Selected area diffraction patterns obtained from the β phase of a Ti-18Zr-14Nb alloy.

V1 variant are expressed as $(0001)_\omega-(111)_\beta$ and $\langle 11\bar{2}0\rangle_\omega-[1\bar{1}0]_\beta$. Fig. 1.30 shows typical SAD patterns obtained from a metastable β-type Ti-based alloy (Ti-18Zr-14Nb) after solution treatment followed by quenching [94]. In the SAD pattern obtained along the $[101]_\beta$ zone axis, the diffraction spots from two variants, i.e., V3 and V4, of the athermal ω phase are distinguished. In this zone axis, other variants (V1 and V2) of the athermal ω phase cannot be identified since the diffraction spots are overlapped with those of the β phase. It is also noted that the intensities of the spots from V3 and V4 are similar. On the other hand, the diffraction spots from the V1 and V2 variants appear with similar intensities in the SAD pattern obtained along the $[102]_\beta$ zone axis. In the SAD pattern obtained along the $[113]_\beta$ zone axis, the diffraction spots from the V2 and V3 variants with similar intensities indicate that four variants of the athermal ω phase are formed equivalently.

1.5.2 Influence on Martensitic Transformation

It has been well acknowledged that the presence of the athermal ω phase hinders the $\beta \rightarrow \alpha''$ martensitic transformation. Fig. 1.31 shows a schematic representation of the effect of the $\beta \rightarrow \omega$ transformation on the martensitic

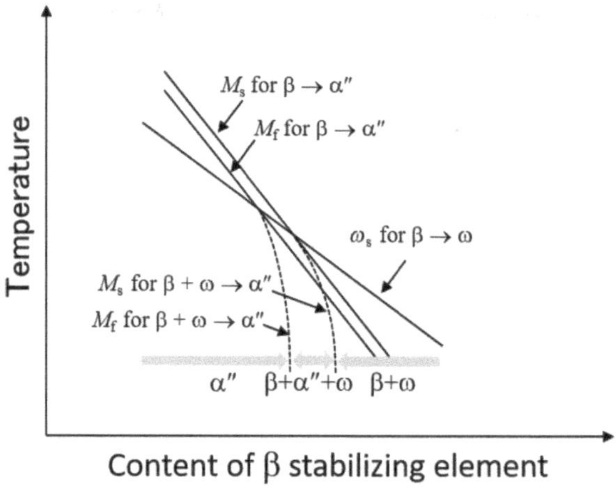

Fig. 1.31 Schematic explanation of the effect of the β → ω transformation on the martensitic transformation temperature and phase constitution.

transformation temperatures. The martensitic transformation start temperature M_s for the β → α″ transformation decreases with the increase of the content of β-stabilizing alloying element. Similarly, the start temperature for the β → ω_{ath} transformation, namely, ω_s, decreases with the increase of the content of β-stabilizing alloying element. Because the slopes of the two lines (M_s and ω_s) are different, they intersect at one point. For the alloys with a higher content of β-stabilizing alloying element, the β → ω_{ath} transformation occurs first upon quenching; therefore, the high-temperature phase for the martensitic transformation becomes a two-phase (β + ω_{ath}) structure, causing a drastic decrease in M_s as indicated by a thick dashed line. The alloys with a relatively lower content of β-stabilizing alloying element exhibit a single phase structure of α″ martensite after quenching to room temperature because both M_s and M_f are higher than ω_s. In the composition range between two dashed lines, the martensitic transformation is not completed, and three phases of β and α″ and ω_{ath} are coexisted.

Not only the slopes of the M_s and ω_s lines but also the composition where two lines intersect is dependent on the kind of alloying elements. As shown in Fig. 1.8, the inclination of the M_s line of the Ti-Mo alloys is the largest among the three kinds of alloys. And the martensite transformation has not been reported in the Ti-Mo alloys with the Mo content above 5 at.%, suggesting that two lines (M_s and ω_s) intersect at much high temperature and low alloy content compared with other alloy systems. Al-Zain et al. [40] also claimed that the addition of Mo gives a stronger effect on ω_s than M_s of

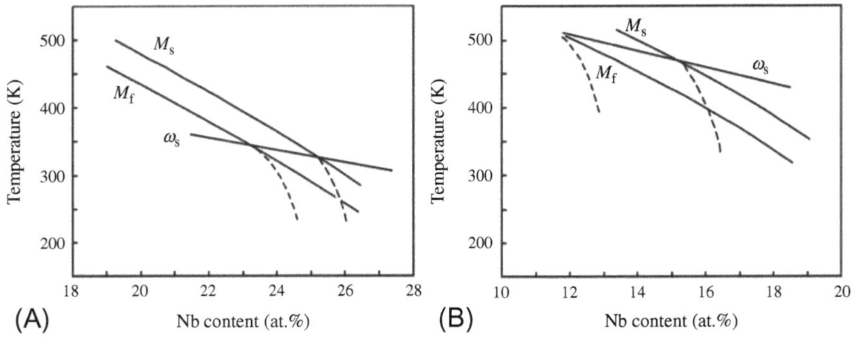

Fig. 1.32 Effect of Nb content on the M_s and ω_s in the (A) Ti-*x*Nb binary alloys and (B) Ti-*x*Nb-2Mo alloys.

Ti-Nb alloys as shown in Fig. 1.32. The addition of 2 at.% Mo relatively raises ω_s with respect to M_s, causing an increase in the temperature where M_s and ω_s lines cross.

1.5.3 Influence on Shape Memory Properties

Both ω_{ath} and ω_{th} phases affect shape memory properties. Fig. 1.33 shows stress-strain curves obtained by a loading-unloading tensile test at various temperatures for Ti-27Nb, Ti-24Nb-1Mo, Ti-21Nb-2Mo, Ti-18Nb-3Mo, and Ti-13Nb-4Mo alloys that have different amount of the ω_{ath} phase [41]. Fig. 1.34 shows SAD patterns of Ti-27Nb, Ti-24Nb-1Mo, Ti-21Nb-2Mo, Ti-18Nb-3Mo, and Ti-13Nb-4Mo alloys along the $[113]_\beta$ zone axis and dark-field images taken from the diffraction spot of the ω phase indicated by a circle [41]. The ω phase is considered as the ω_{ath} phase because the alloys are subjected to the solution treatment followed by quenching. It is clearly seen that the intensity of diffraction spots of the ω_{ath} phase became stronger with increasing Mo content or with decreasing Nb content. The dark-field images also show that the volume fraction of the ω_{ath} phase increases with increasing Mo content or with decreasing Nb content.

Fig. 1.33 shows that the temperature dependence of shape memory behavior is strongly influenced by the composition although all alloys exhibit superelasticity at room temperature (298 K). For the Ti-27Nb alloy, superelasticity was observed in the temperature range 253–298 K. At lower temperatures 193–233 K, shape memory effect appeared; the remained strain is recovered by heating the specimen. It is noted that the apparent yield stress, i.e., the critical stress for onset of the stress-induced martensitic transformation, decreases with decreasing test temperature. This temperature

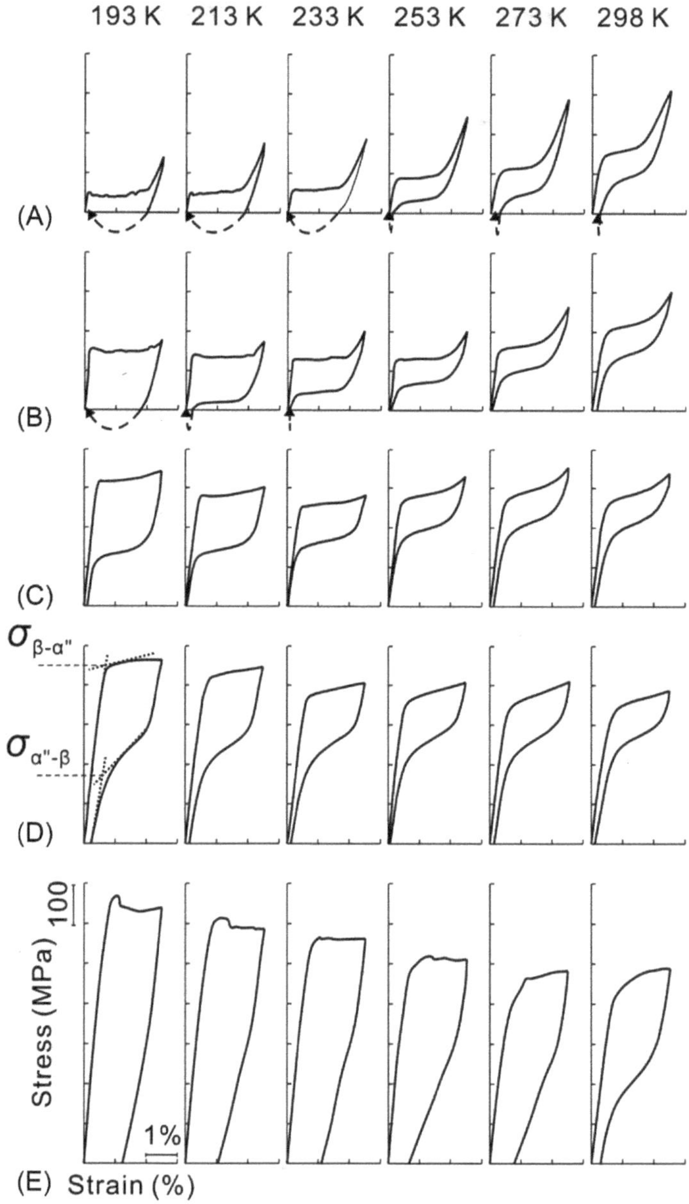

Fig. 1.33 Stress-strain curves at various temperature for (A) Ti-27Nb, (B) Ti-24Nb-1Mo, (C) Ti-21Nb-2Mo, (D) Ti-18Nb-3Mo, and (E) Ti-13Nb-4Mo alloys.

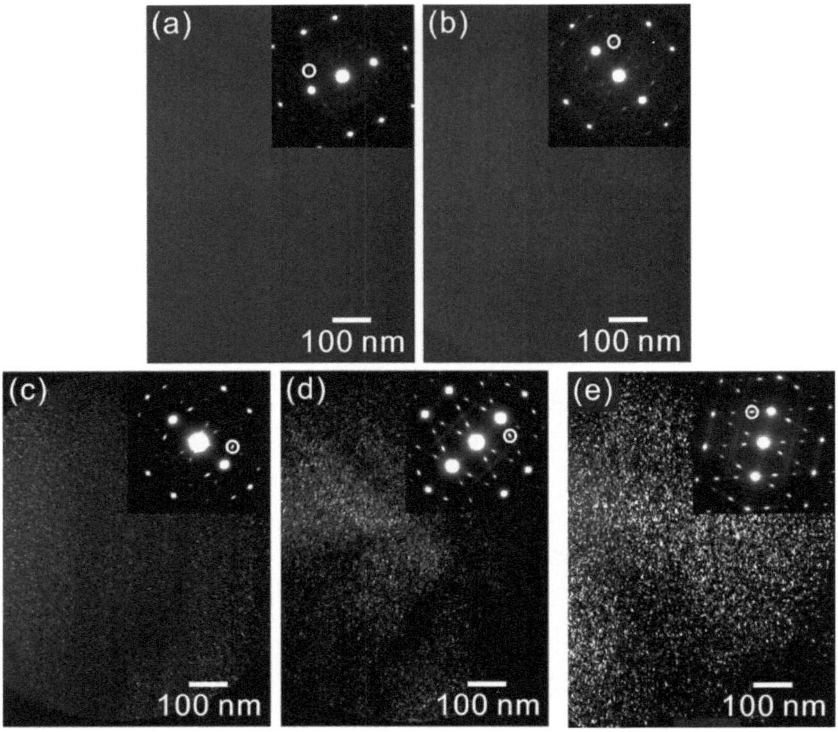

Fig. 1.34 Selected area diffraction patterns and dark-field images taken from the diffraction spot of the ω phase indicated by a circle for (A) Ti-27Nb, (B) Ti-24Nb-1Mo, (C) Ti-21Nb-2Mo, (D) Ti-18Nb-3Mo, and (E) Ti-13Nb-4Mo alloys.

dependence is consistent with the Clausius-Clapeyron relationship between the stress for inducing martensitic transformation (σ) and temperature (T) expressed as follows: $d\sigma/dT = -\Delta H/T \cdot \varepsilon = -\Delta S/\varepsilon$, where ε is a transformation strain and ΔH and ΔS are transformation enthalpy and entropy per unit volume, respectively. In the relationship, the slope of $d\sigma/dT$ is a positive because ΔH and ΔS are negative values. In other words, the martensite phase (α'') becomes more stable, and then, the stress required to induce martensitic transformation decreases with decreasing temperature.

The Ti-24Nb-1Mo alloy exhibits superelasticity in the temperature range 213–298K. It is seen that the critical stress for onset of the stress-induced martensitic transformation decreases from 298 to 253K, but it gradually increases with further decreasing temperature. For the Ti-21Nb-2Mo alloy, the critical stress for onset of the stress-induced martensitic transformation remains almost constant with decreasing temperature from 298 to

233 K. Below 233 K, the critical stress increases with decreasing temperature. In the Ti-18Nb-3Mo and Ti-13Nb-4Mo alloys, the critical stress increases monotonically with decreasing temperature that is the opposite tendency predicted by the Clausius-Clapeyron relationship. It is also worth to note that the stress for the reverse transformation decreases monotonically with decreasing temperature, resulting in the stress hysteresis, i.e., the difference between the stress for forward transformation and that for the reverse transformation, increases with decreasing temperature.

Not only the opposite temperature dependence of the critical stress for onset of the stress-induced martensitic transformation but also the large stress hysteresis has shown to be due to the formation of the ω_{ath} phase. Electric resistivity has been employed as an indicator of phase stability. Fig. 1.35

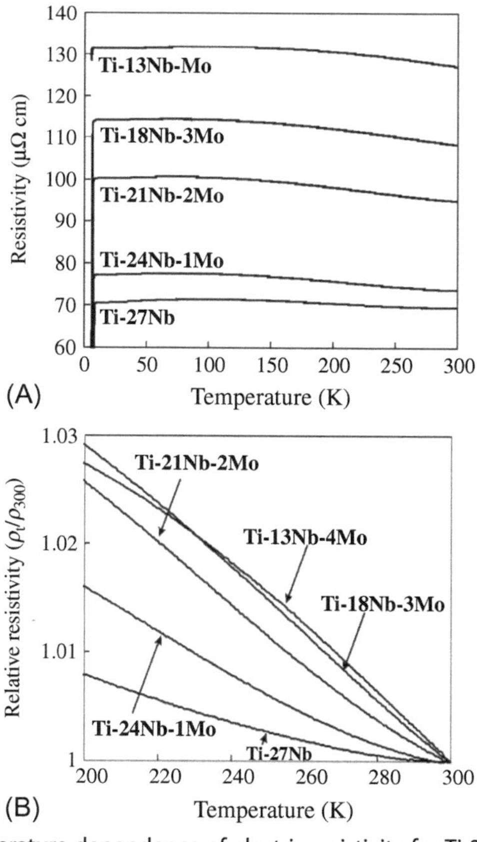

Fig. 1.35 (A) Temperature dependence of electric resistivity for Ti-27Nb, Ti-24Nb-1Mo, Ti-21Nb-2Mo, Ti-18Nb-3Mo, and Ti-13Nb-4Mo alloys. (B) Relative resistivity normalized by the value at 300 K.

shows the temperature dependence of electric resistivity measured upon cooling and heating between 300 and 5 K for the Ti-27Nb, Ti-24Nb-1Mo, Ti-21Nb-2Mo, Ti-18Nb-3Mo, and Ti-13Nb-4Mo alloys [41]. It is noticed that the electric resistivity is higher in the alloy containing higher amount of the ω_{ath} phase. All alloys exhibit an anomalous increase in electric resistivity with decreasing temperature. The negative temperature coefficient of resistivity is related to the increase of the volume fraction of the ω_{ath} phase with decreasing temperature, most possibly due to electric scattering at ω/β interfaces. The increase in the volume fraction of the ω_{ath} phase is directly confirmed by TEM observation. Fig. 1.36 shows SAD patterns of the Ti-13Nb-4Mo alloy obtained at 293 and 90 K and dark-field images taken from the diffraction spot of the ω phase [41]. It is apparently noticed that the volume fraction of the ω_{ath} phase increases with decreasing temperature. As seen in the normalized resistivity profile (Fig. 1.35B), i.e., the relative resistivity normalized by the electric resistivity at 300 K, the rate of the increase in the resistivity is higher in the alloys with a higher Mo content, implying that the anomalous temperature dependence of the superelastic behavior in the Ti-Nb-Mo alloys is closely related to the ω_{ath} phase. Namely, the anomalous increase in the critical stress for onset of the stress-induced martensitic transformation with decreasing temperature is due to the increase in the volume fraction of the ω_{ath} phase.

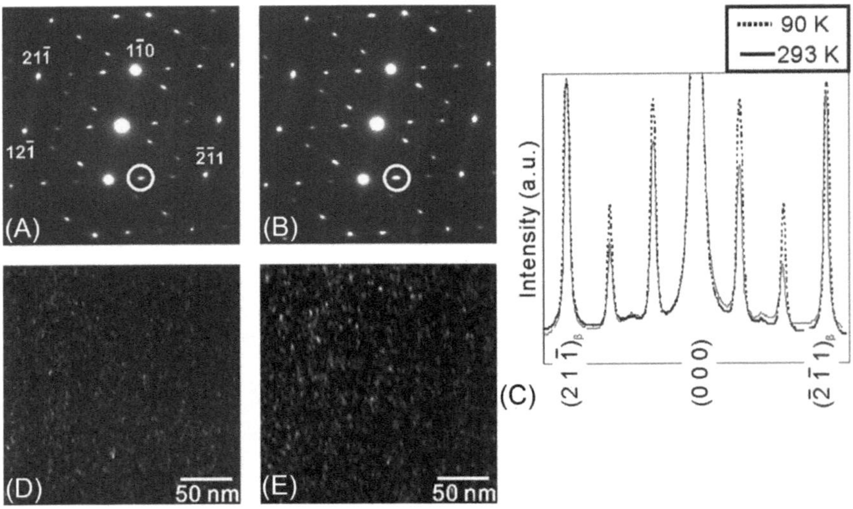

Fig. 1.36 Selected area diffraction patterns of the Ti-13Nb-4Mo alloy obtained at (A) 293 and (B) 90 K. Dark-field images taken from the diffraction spot of the ω phase obtained at (D) 293 and (E) 90 K. (C) Intensity profile of β and ω diffractions.

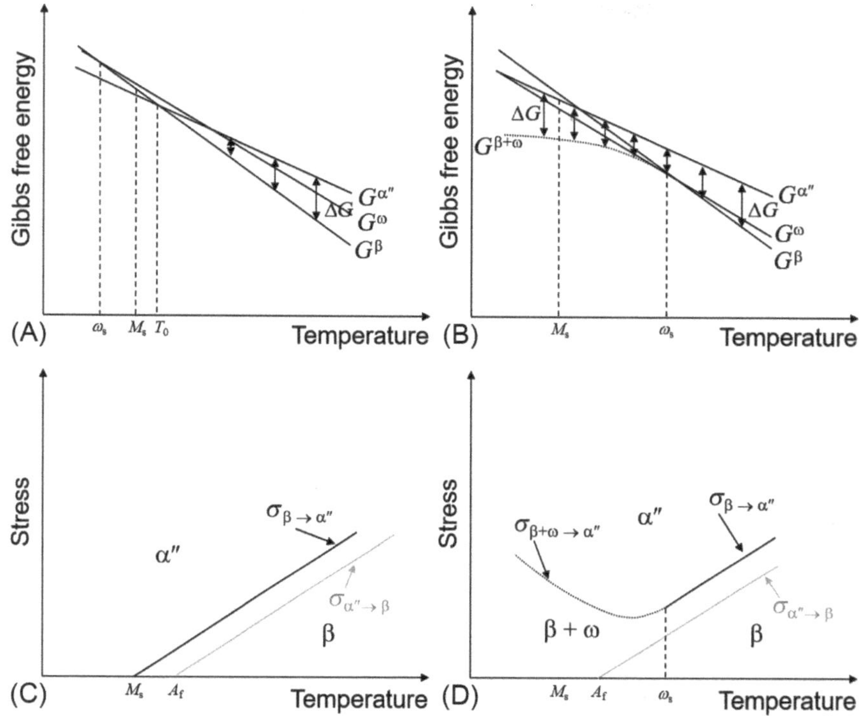

Fig. 1.37 Temperature dependence of Gibbs free energies of three phases (β, α″, and ω_{ath}) in the case that (A) M_s is higher than ω_s and (B) ω_s is higher than M_s. (C) and (D) are the corresponding temperature dependence of the stress-induced martensitic transformation (β → α″) of the (A) and (B), respectively.

Anomalous superelastic behavior can be understood using the schematic shown in Fig. 1.37. Fig. 1.37A describes the temperature dependence of Gibbs free energies of three phases (β, α″, and ω_{ath}) where M_s is higher than ω_s. In this situation, the effect of the ω_{ath} phase on the stress-induced martensitic transformation (β → α″) can be ignored. The difference of Gibbs free energy ΔG $(= G^{\alpha''} - G^{\beta})$ decreases with decreasing temperature, and then, the stress for inducing martensitic transformation decreases with decreasing temperature, as with conventional shape memory alloys, as shown in Fig. 1.37C. If ω_s is higher than M_s as shown in Fig. 1.37B, the transformation can be divided into two groups. At higher temperatures above ω_s, the stress-induced martensitic transformation occurs from β to α″. On the other hand, the α″ phase is stress-induced from a two-phase (β + ω_{ath}) structure at lower temperatures below ω_s; therefore, the Gibbs free energy difference for the stress-induced martensitic transformation is changed from $G^{\alpha''} - G^{\beta}$

to $G^{\alpha''} - G^{\beta+\omega}$. As a result, the magnitude of the Gibbs free energy difference decreases with decreasing temperature until reaching the minimum value at a temperature below ω_s, and then, it increases with further decreasing temperature. Accordingly, the stress for inducing martensitic transformation decreases linearly with decreasing temperature until ω_s, and it starts to deviate from the linear relationship from ω_s as shown in Fig. 1.37D. After reaching the minimum value, the stress for inducing martensitic transformation increases with further decreasing temperature. It has been claimed that the ω_{ath} phase transforms into the α'' martensite phase [38, 41, 47]. Al-Zain et al. [41] reported an in situ XRD experiment showing that the ω_{ath} phase transforms into the α'' structure during loading. Inamura et al. [38] reported that the ω_{ath} phase transforms into the α'' structure through shearing and rearrangement of atoms although stacking faults remain. Therefore, it is reasonable to expect that a higher stress is required to initiate the stress-induced martensitic transformation. Consequently, we believe that this model is plausible and well describes the abnormal experimental results.

It was reported that the thermal ω phase (ω_{th}) formed by aging causes the anomalous temperature dependence of superelastic behavior of the Ti-27Nb alloy. Fig. 1.38 shows stress-strain curves obtained by a loading-unloading tensile test at various temperatures for the as-annealed and aged specimens of the Ti-27Nb alloy [95]. In the as-annealed specimen (AG) and the specimen aged at 773 K, the critical stress for onset of the stress-induced martensitic transformation decreases with decreasing test temperature. On the other hand, in the specimens aged at 573 and 623 K, the critical stress for onset of the stress-induced martensitic transformation decreases gradually with decreasing temperature, reaching a minimum value, and then increases with further decreasing temperature. Al-Zain et al. [95] claimed that the ω_{th} precipitates provide nucleation sites and then encourages the formation of the ω_{ath} phase upon cooling.

1.5.4 Room Temperature Aging

It has been considered that the formation of the ω_{ath} is essentially diffusionless and displacive in nature. However, the room-temperature aging effect has been confirmed to occur in metastable β-type Ti-based alloys. Tane et al. [96] reported that the electric resistivity and shear moduli c' and c_{44} are increased by room-temperature aging in Ti-Nb-Ta-Zr and Ti-Nb-Ta-Zr-O alloys. They proposed that the increase in the electric

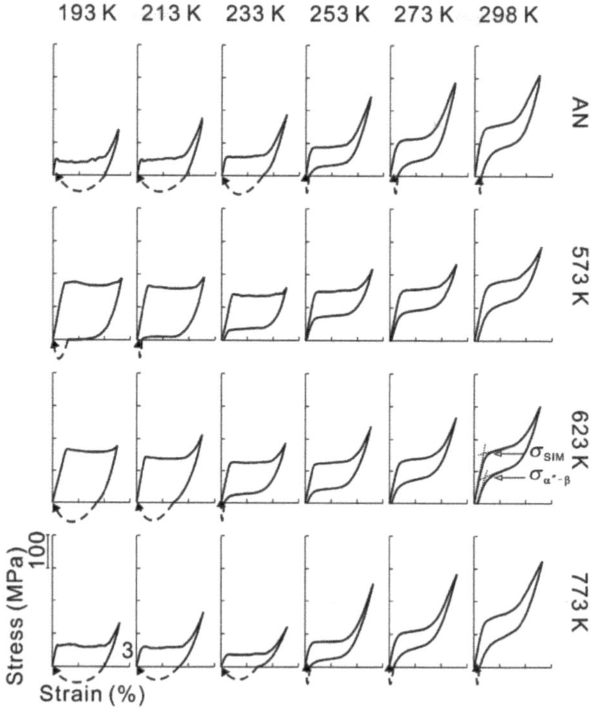

Fig. 1.38 Stress-strain curves obtained at various temperatures for the as-annealed and aged specimens of the Ti-27Nb alloy.

resistivity and shear moduli is due to the formation of a diffuse ω structure. Al-Zain et al. [42] reported that the stress for inducing martensitic transformation is affected by room-temperature aging. Fig. 1.39 shows the effect of room-temperature aging (28 days) on the superelastic behavior of Ti-18Nb-3Mo, Ti-16Nb-3Mo-1Sn, and Ti-13.5Nb-3Mo-2Sn alloys. The stress-strain curves noted as $n=1$ were obtained by an loading-unloading tensile test at room temperature immediately after annealing at 973 K for 0.6 ks, whereas the results noted as $n'=1$ were obtained after aging at room temperature for 28 days subsequent to annealing. The results show that the onset stress for inducing martensitic transformation is increased by room-temperature aging. It is also seen that the increment of the onset stress in the Ti-18Nb-3Mo is much higher when compared with the Sn-containing alloy. Fig. 1.40 shows selected area diffraction patterns and dark-field images of the ω phase of the Ti-18Nb-3Mo alloy taken before room-temperature aging (Fig. 1.40A and B) and after room-temperature aging for 28 days (Fig. 1.40A′ and B′), respectively [42].

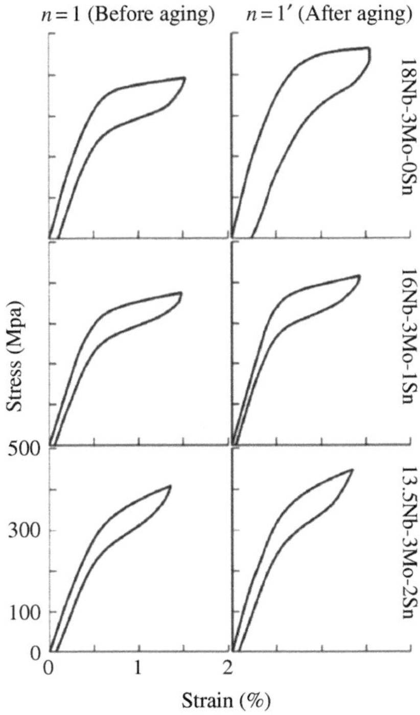

Fig. 1.39 Effect of room-temperature aging (28 days) on the stress-strain curves of Ti-18Nb-3Mo, Ti-16Nb-3Mo-1Sn, and Ti-13.5Nb-3Mo-2Sn alloys.

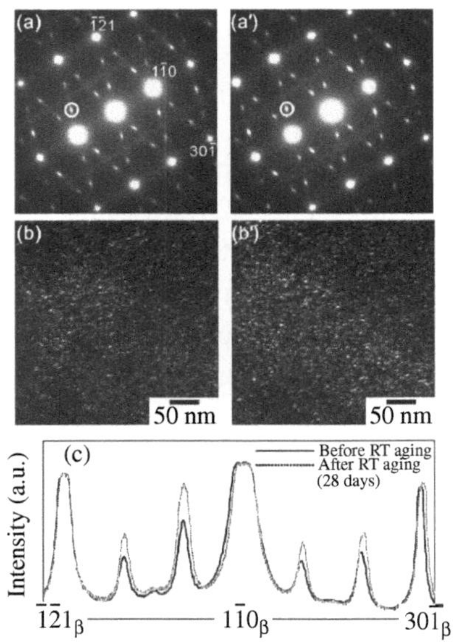

Fig. 1.40 Effect of room-temperature aging (28 days) on the ω phase in the Ti-18Nb-3Mo alloy. Selected area diffraction patterns taken (A) before room-temperature aging and (B) after room-temperature aging. (A′) and (B′) are dark-field images taken from the diffraction spot of the ω phase in (A) and (B), respectively. (C) Intensity profile of β and ω diffractions.

It is clearly seen from Fig. 1.40C that the intensity of the diffraction spots from the ω phase became strong after room-temperature aging. Dark-field images also show that the volume fraction of the ω phase increased by room-temperature aging. The increase in the onset stress of martensitic transformation leads to detrimental results on the superelastic properties as shown in Fig. 1.39. The addition of Sn is effective to reduce the room-temperature aging effect by suppressing the formation of the ω phase, resulting in much stable superelastic properties.

1.6 SUMMARY

As a result of continued research, many fundamental aspects of the $\beta \rightarrow \alpha''$ martensitic transformation in Ti alloys, such as transformation temperature, crystal structure of martensite phase, habit planes, substructure of martensite, and self-accommodation morphology, have been clarified. The α'' martensite phase has an orthorhombic structure that is a compromise structure or an intermediate phase between the β phase with a bcc structure and the α' phase with a hexagonal structure. The orthorhombicity decreases with increasing amount of β-stabilizing alloying element such as Nb, Ta, and Mo. Although there have been many reports on the martensitic transformation behavior with some reproducibility in the Ti-Nb alloys, there have been comparatively few studies in other alloy system (Ti-Ta and Ti-Mo). The martensitic transformation temperature decreases almost linearly with increasing content of alloying element. Only limited quantitative results on the martensitic transformation temperature in multicomponent alloys have been reported. Much uncertainty still remains about the effect of alloying element on the crystal structure and martensitic transformation temperature. The crystallography of martensite, i.e., habit plane and self-accommodation morphology, has been well explained by the phenomenological theory of martensite crystallography. It has been confirmed that the martensite phases are disordered in β-type Ti-based alloys, which is unusual comparing with other conventional shape memory alloys. The unique microstructure of APB-like domain contrast has been confirmed to be due to the displacement induced fault caused by two different modes of shuffling. Understanding on the $\beta \rightarrow \omega$ transformation is indispensable because it exerts a strong effect on the $\beta \rightarrow \alpha''$ martensitic transformation and shape memory effect.

REFERENCES

[1] P. Duwez, Effect of rate of cooling on the alpha–beta transformation in titanium and titanium–molybdenum alloys, JOM 3 (1951) 765–771.
[2] Y.C. Liu, H. Margolin, Martensite habit plane in quenched Ti–Mn alloys, JOM 5 (1953) 667–670.
[3] P. Duwez, The martensitic transformation temperature in titanium binary alloys, Trans. ASM 45 (1953) 934–940.
[4] Y.C. Liu, Martensitic transformation in binary titanium alloys, JOM 8 (1956) 1036–1040.
[5] I.A. Bagariatskii, G.I. Nosova, T.V. Tagunova, Factors in the formation of metastable phases in titanium–base alloys, Sov. Phys. Doklady 3 (1958) 1014–1018.
[6] P. Gaunt, J.W. Christian, The crystallography of the β–α transformation in zirconium and in two titanium–molybdenum alloys, Acta Metall. 7 (1959) 534–543.
[7] H. Kaneko, Y.C. Huang, Allotropic transformation characteristics of titanium alloys during continuous cooling, J. Japan Inst. Metals 27 (1963) 387–393.
[8] A.R.G. Brown, D. Clark, J. Eastabrook, K.S. Jepson, Titanium–niobium system, Nature 201 (1964) 914–915.
[9] K.S. Jepson, A.R.G. Brown, J.A. Gray, in: R.I. Jaffee, N.E. Promisel (Eds.), The effect of cooling rate on the beta transformation in titanium–niobium and titanium–aluminum alloys, The Science, Technology and Applications of Titanium: Proceedings of 1st International Conference on Titanium, Pergamon Press, Oxford, 1970, , pp. 677–690.
[10] Y.C. Huang, S. Suzuki, H. Kaneko, T. Sato, in: R.I. Jaffee, N.E. Promisel (Eds.), Thermodynamics of the M_s points in titanium alloys, The Science, Technology and Applications of Titanium: Proceedings of 1st International Conference on Titanium, Pergamon Press, Oxford, 1970, , pp. 691–698.
[11] C. Baker, The shape–memory effect in a titanium–35 wt.% niobium alloy, J. Metal Sci. 5 (1971) 92–100.
[12] D.L. Moffat, D.C. Larbalestier, The competition between martensite and omega in quenched Ti–Nb alloys, Metall. Trans. 19 (1988) 1677–1686.
[13] J. Zhao, M.R. Notis, Phase transformation kinetics and the assessment of equilibrium and metastable states, J. Phase Equil. 14 (1993) 304–315.
[14] A. Dobromyslov, V. Elkin, Martensitic transformation and metastable beta–phase in binary titanium alloys with d-metals of 4–6 periods, Scr. Mater. 44 (2001) 905–910.
[15] H. Matsumoto, S. Watanabe, S. Hanada, Beta TiNbSn alloys with low Young's modulus and high strength, Mater. Trans. 46 (2005) 1070–1078.
[16] H.Y. Kim, Y. Ikehara, J.I. Kim, H. Hosoda, S. Miyazaki, Martensitic transformation, shape memory effect and superelasticity of Ti–Nb binary alloys, Acta Mater. 54 (2006) 2419–2429.
[17] S. Neelakantan, P.E.J. Rivera-Diaz-del- Castillo, S. van der Zwaag, Prediction of the martensite start temperature for β titanium alloys as a function of composition, Scripta Mater. 60 (2009) 611–614.
[18] R.R. Hake, D.H. Leslie, T.G. Berlincourt, Electrical resistivity, Hall effect and superconductivity of some b.c.c. titanium–molybdenum alloys, J. Phys. Chem. Solids 20 (1961) 177–186.
[19] B.A. Hatt, V.G. Rivlin, Phase transformations in superconducting Ti–Nb alloys, J. Phys. D 1 (1968) 1145–1149.
[20] C. Baker, J. Sutton, Correlation of superconducting and metallurgical properties of a Ti–20 at.% Nb alloy, Phil. Mag. 19 (1969) 1223–1255.
[21] R. Davis, H.M. Flower, D.R.F. West, Martensitic transformations in Ti–Mo alloys, J. Mater. Sci. 14 (1979) 712–722.

[22] T. Ahmed, H.J. Rack, Martensitic transformations in Ti–(16–26 at %) Nb alloys, J. Mater. Sci. 31 (1996) 4267–4276.
[23] A.V. Dobromyslov, V.A. Elkin, The orthorhombic α″–phase in binary titanium–base alloys with d–metals of V–VIII groups, Mater. Sci. Eng. A 438 (2006) 324–326.
[24] N.B. D'yakonova, I.V. Lyasotskii, Y.L. Rodionov, Orthorhombic martensite and the ω phase in quenched and deformed titanium alloys with 20–24 at.% Nb, Russ. Metall. Met. 2007 (2007) 61–70.
[25] S. Banumathy, R.K. Mandal, A.K. Singh, Structure of orthorhombic martensitic phase in binary Ti–Nb alloys, J. Appl. Phys. 106 (2009). 093518.
[26] Y. Takemoto, Y. Miyake, T. Senuma, Microstructure and mechanical properties of Ti–Mo and Ti–6Mo–X–Y alloys, J. Japan Inst. Metals 73 (2009) 752–757.
[27] R. Hermann, H. Hermann, M. Calin, B. Buechner, J. Eckert, Elastic constants of single crystalline beta–$Ti_{70}Nb_{30}$, Scr. Mater. 66 (2012) 198–201.
[28] M. Bonisch, M. Calin, L. Giebeler, A. Helth, A. Gebert, W. Skrotzki, J. Eckert, Composition–dependent magnitude of atomic shuffles in Ti–Nb martensites, J. Appl. Crystallogr. 47 (2014) 1374–1379.
[29] J.Y. Yan, G.B. Olson, Molar volumes of bcc, hcp, and orthorhombic Ti–base solid solutions at room temperature, Calphad 52 (2016) 152–158.
[30] M. Sabeena, S. Murugesan, P. Anees, E. Mohandas, M. Vijayalakshmi, Crystal structure and bonding characteristics of transformation products of bcc β in Ti–Mo alloys, J. Alloys Compd. 705 (2017) 769–781.
[31] K.A. Bywater, J.W. Christian, Martensitic transformations in titanium–tantalum alloys, Phil. Mag. 25 (1972) 1249–1273.
[32] H.M. Flower, S.D. Henry, D.R.F. West, The β–α transformation in dilute Ti–Mo alloys, J. Mater. Sci. 9 (1974) 57–64.
[33] J.S. Lee Pak, C.Y. Lei, C.M. Wayman, Atomic ordering in Ti–V–A1 shape memory alloys, Mater. Sci. Eng. A132 (1991) 237–244.
[34] D. Banerjee, K. Muraleedharan, J.L. Strudel, Substructure in titanium alloy martensite, Phil. Mag. A 77 (1998) 299–323.
[35] T. Inamura, J.I. Kim, H.Y. Kim, H. Hosoda, K. Wakashima, S. Miyazaki, Composition dependent crystallography of α″-martensite in Ti–Nb-based β-titanium alloy, Phil. Mag. 87 (2007) 3325–3350.
[36] Y.W. Chai, H.Y. Kim, H. Hosoda, S. Miyazaki, Interfacial defects in Ti–Nb shape memory alloys, Acta Mater. 56 (2008) 3088–3097.
[37] Y.W. Chai, H.Y. Kim, H. Hosoda, S. Miyazaki, Self-accommodation in Ti–Nb shape memory alloys, Acta Mater. 57 (2009) 4054–4064.
[38] T. Inamura, H. Hosoda, H.Y. Kim, S. Miyazaki, Antiphase boundary-like stacking fault in α″-martensite of disordered crystal structure in β-titanium shape memory alloy, Phil. Mag. 90 (2010) 3475–3498.
[39] S. Banerjee, P. Mukhopadhyay, Phase Transformations: Examples From Titanium and Zirconium Alloys, Elsevier, Amsterdam, 2010.
[40] Y. Al-Zain, H.Y. Kim, H. Hosoda, T.H. Nam, S. Miyazaki, Shape memory properties of Ti–Nb–Mo biomedical alloys, Acta Mater. 58 (2010) 4212–4223.
[41] Y. Al-Zain, H.Y. Kim, T. Koyano, H. Hosoda, T.H. Nam, S. Miyazaki, Anomalous temperature dependence of the superelastic behavior of Ti–Nb–Mo alloys, Acta Mater. 59 (2011) 1464–1473.
[42] Y. Al-Zain, Y. Sato, H.Y. Kim, H. Hosoda, T.H. Nam, S. Miyazaki, Room temperature aging behavior of Ti–Nb–Mo-based superelastic alloys, Acta Mater. 60 (2012) 2437–2447.
[43] D.S. Lieberman, M.S. Wechsler, T.A. Read, Cubic to orthorhombic diffusionless phase change—experimental and theoretical studies of AuCd, J. Appl. Phys. 26 (1955) 473–484.

[44] A.E. Dwight, R.A. Conner Jnr, J.W. Downey, Equiatomic compounds of the transition and lanthanide elements with Rh, Ir, Ni and Pt, Acta Cryst. 18 (1965) 835–839.
[45] T.H. Nam, T. Saburi, K. Shimizu, Cu-content dependence of shape memory characteristics in Ti–Ni–Cu alloys, Trans. JIM 31 (1990) 959–967.
[46] M.J. Duggin, W.A. Raghinger, The nature of the martensite transformation in a copper–nickel–aluminium alloy, Acta Metall. 12 (1964) 529–535.
[47] T.W. Duerig, J. Albrecht, D. Richter, P. Fischer, Formation and reversion of stress induced martensite in Ti–10V–2Fe–3Al, Acta Metall. 30 (1982) 2161–2172.
[48] S.G. Fedotov, T.V. Chelidze, Y.K. Kovneristyi, V.V. Sanadze, Phase structure, critical points *Ms* and *As* of martensitic transformation and elastic properties of metastable alloys of the Ti–Ta system, Phys. Met. Metallogr. 60 (1985) 139–143.
[49] S.G. Fedotov, T.V. Chelidze, Y.K. Kovneristyi, K. Yu, V.V. Sanadze, Phase transformations in metastable alloys of Ti–Ta system during heating, Phys. Met. Metallogr. 62 (1986) 328–332.
[50] A.V. Dobromyslov, G.V. Dolgikh, Y. Dutkevich, T.L. Trenogina, Phase and structural transformations in Ti–Ta alloys, Phys. Met. Metallogr. 107 (2009) 502–510.
[51] P.J.S. Buenconsejo, H.Y. Kim, S. Miyazaki, Shape memory behavior of Ti–Ta and its potential as a high-temperature shape memory alloy, Acta Mater. 57 (2009) 1068–1077.
[52] D.E. Huber, Structure and Properties of Titanium Tantalum Alloys for Biocompatibility, PhD Thesis, The Ohio State University, 2016.
[53] C. Leibovitch, E. Gartstein, A.G. Rabinkin, Structural stability and superconductivity of Ti–Mo alloys under pressure. Part I. Structural stability, Z. Metallk. 71 (1980) 438–447.
[54] K. Otsuka, X. Ren, Physical metallurgy of Ti–Ni-based shape memory alloys, Prog. Mater. Sci. 50 (2005) 511–678.
[55] H.C. Donkersloot, J.H.N. Van Vucht, Martensitic transformations in gold–titanium, palladium-titanium, platinum-titanium alloys near the equiatomic composition, J. Less-Common Metals 20 (1970) 83–91.
[56] Y. Murakami, N. Nakanishi, S. Kachi, Electron diffraction study on the structure of the Au-Cu-Zn martensites, Jpn. J. Appl. Phys. 11 (1972) 1591–1598.
[57] J. Pons, V.A. Chernenko, R. Santamaria, E. Cesari, Crystal structure of martensitic phases in Ni–Mn–Ga shape memory alloys, Acta Mater. 48 (2000) 3027–3038.
[58] D. Schryvers, Martensitic and related transformations in Ni–Al alloys, J. Phys. Colloq. 05 (C2) (1995) 225–234.
[59] K. Ando, T. Omori, I. Ohnuma, R. Kainuma, K. Ishida, Ferromagnetic to weak-magnetic transition accompanied by bcc to fcc transformation in Fe–Mn–Al alloy, Appl. Phys. Lett. 95 (2009). 212504.
[60] C. Zener, Contributions to the theory of beta-phase alloys, Phys. Rev. 71 (1947) 846–851.
[61] N. Nakanishi, Elastic constants as they relate to lattice properties and martensite formation, Prog. Mater. Sci. 24 (1980) 143–265.
[62] B. Verliden, L. Delaey, On the elastic constants and Ms-temperatures in β-hume-rothery alloys, Acta Metall. 36 (1988) 1771–1779.
[63] X. Ren, N. Miura, J. Zhang, K. Otsuka, K. Tanaka, M. Koiwa, T. Suzuki, Y. I. Chumlyakov, M. Asai, A comparative study of elastic constants of Ti–Ni based alloys prior to martensitic transformation, Mater. Sci. Eng. A 312 (2001) 196–206.
[64] H.W. Jeong, Y.S. Yoo, Y.T. Lee, J.K. Park, Elastic softening behavior of Ti–Nb single crystal near martensitic transformation temperature, J. Appl. Phys. 108 (2010). 063515.
[65] Y.P. Hou, S. Guoa, X.L. Qiao, T. Tian, Q.K. Meng, X.N. Cheng, X.Q. Zhao, Origin of ultralow Young's modulus in a metastable β-type Ti–33Nb–4Sn alloy, J. Mech. Behav. Biomed. Mater. 59 (2016) 220–225.

[66] Y.W. Zhang, S.J. Li, E.G. Obbard, H. Wang, S.C. Wang, Y.L. Hao, R. Yang, Elastic properties of Ti–24Nb–4Zr–8Sn single crystals with bcc crystal structure, Acta Mater. 59 (2011) 3081–3090.
[67] O. Mercier, K.N. Melton, G. Gremaud, J. Hagi, Single-crystal elastic constants of the equiatomic NiTi alloy near the martensitic transformation, J. Appl. Phys. 51 (1980) 1833–1834.
[68] V.N. Khachin, S.A. Muslov, V.G. Pushin, Y.I. Chumlyakov, Anomalies of elastic properties of TiNi-TiFe single crystals, Sov. Phys. Dokl. 32 (1987) 606–609.
[69] T.M. Brill, S. Mittelbach, W. Assmus, M. Mullner, B. Luthi, Elastic properties of NiTi, J. Phys. Condens. Matter. 3 (1991) 9621–9627.
[70] G. Guenin, M. Morin, P.F. Gobin, W. Dijonghe, L. Delaey, Elastic constant measurements in β Cu–Zn–Al near the martensitic transformation temperature, Scripta Metall. 11 (1977) 1071–1075.
[71] S. Zirinsky, The temperature dependence of the elastic constants of gold–cadmium alloys, Acta Metall. 4 (1956) 164–171.
[72] K. Enami, J. Hasunuma, A. Nagasawa, S. Nenno, Elastic softening and electron-diffraction anomalies prior to the martensitic transformation in a Ni–Al β alloy, Scripta Metall. 10 (1976) 879–884.
[73] Y. Fukui, T. Inamura, H. Hosoda, K. Wakashima, S. Miyazaki, Mechanical properties of a Ti–Nb–Al shape memory alloy, Mater. Trans. 45 (2004) 1077–1082.
[74] P.J.S. Buenconsejo, H.Y. Kim, S. Miyazaki, Effect of ternary alloying elements on the shape memory behavior of Ti–Ta alloys, Acta Mater. 57 (2009) 2509–2515.
[75] E. Takahashi, T. Sakurai, S. Watanabe, N. Masahashi, S. Hanada, Effect of heat treatment and Sn content on superelasticity in biocompatible TiNbSn alloys, Mater. Trans. 43 (2002) 2978–2983.
[76] J.I. Kim, H.Y. Kim, T. Inamura, H. Hosoda, S. Miyazaki, Shape memory characteristics of Ti–22Nb–(2-8)Zr(at.%) biomedical alloys, Mater. Sci. Eng. A 403 (2005) 334–339.
[77] H.Y. Kim, K. Nakai, J. Fu, S. Miyazaki, Effect of Al addition on superelastic properties of Ti–Zr–Nb-based alloys, Funct. Mater. Lett. 10 (2017). 1740002.
[78] S.M. Tan, S. Miyazaki, Ti-content and annealing temperature dependence of deformation characteristics of $Ti_xNi_{(92-x)}Cu_8$ shape memory alloys, Acta Mater. 46 (1997) 2729–2740.
[79] S. Miyazaki, V.H. No, K. Kitamura, K. Anak, H. Hosoda, Texture of Ti–Ni rolled thin plates and sputter-deposited thin films, Inter. J. Plast. 16 (2000) 1135–1154.
[80] M.S. Wechsler, D.S. Lieberman, T.A. Read, On the theory of the formation of martensite, Trans. AIME 197 (1953) 1503–1515.
[81] J.S. Bowles, J.K. Mackenzie, The crystallography of martensite transformations I, Acta Metall. 2 (1954) 129–137.
[82] J.S. Bowles, J.K. Mackenzie, The crystallography of martensite transformations II, Acta Metall. 2 (1954) 138–147.
[83] D.S. Lieberman, M.S. Wechsler, T.A. Read, Cubic to orthorhombic diffusionless phase change—experimental and theoretical studies of AuCd, J. Appl. Phys. 26 (1955) 473–489.
[84] M. Tahara, N. Okano, T. Inamura, H. Hosoda, Plastic deformation behavior of single-crystalline martensite of Ti–Nb shape memory alloy, Sci. Rep. 7 (2017). 15715.
[85] O.M. Ivasishin, N.S. Kosenko, S.V. Shevchenko, Crystallographic features of α″-martensite in titanium alloys, J. Phys. IV 05 (C8) (1995) 1017–1022.
[86] R.C. Pond, S. Celotto, J.P. Hirth, A comparison of the phenomenological theory of martensitic transformations with a model based on interfacial defects, Acta Mater. 51 (2003) 5385–5398.

[87] S. Miyazaki, K. Otsuka, C.M. Wayman, The shape memory mechanism associated with the martensitic transformation in Ti–Ni alloys—I. Self-accommodation, Acta Metall. 37 (1989) 1873–1884.
[88] K.M. Knowles, A high resolution electron microscope study of martensite and martensitic interfaces in titanium-manganese, Proc. R. Soc. Ser. A 380 (1982) 187–200.
[89] M. Matsuda, T. Hara, M. Nishida, Crystallography and morphology of antiphase boundary-like structure induced by martensitic transformation in Ti-Pd shape memory alloy, Mater. Trans. 49 (2008) 461–465.
[90] B.S. Hickman, The formation of omega phase in titanium and zirconium alloys: a review, J. Mater. Sci. 4 (1969) 554–563.
[91] D. De Fontaine, N.E. Paton, J.C. Williams, The omega phase transformation in titanium alloys as an example of displacement controlled reactions, Acta Metall. 19 (1971) 1153–1162.
[92] E.W. Collings, The Physical Metallurgy of Titanium Alloys, American Society of Metals, Metals Park, OH, 1984.
[93] S.K. Sikka, Y.K. Vohra, R. Chidambaram, Omega phase in materials, Prog. Mater. Sci. 27 (1982) 245–310.
[94] H.Y. Kim, K. Nakai, S. Miyazaki, Unpublished work.
[95] Y. Al-Zain, H.Y. Kim, T. Koyano, H. Hosoda, S. Miyazaki, A comparative study on the effects of the ω and α phases on the temperature dependence of shape memory behavior of a Ti–27Nb alloy, Scripta Mater. 103 (2015) 37–40.
[96] M. Tane, K. Hagihara, M. Ueda, T. Nakano, Y. Okuda, Elastic-modulus enhancement during room-temperature aging and its suppression in metastable Ti–Nb-based alloys with low body-centered cubic phase stability, Acta Mater. 102 (2016) 373–384.

CHAPTER 2

Shape Memory Effect and Superelasticity

Contents

2.1 INTRODUCTION

The shape memory effect in Ti-based alloys was first observed by Baker in 1971 in a Ti-35 wt% Nb alloy [1]. In the 1980s, the stress-induced martensitic transformation from β phase to α″ phase and shape memory effect were reported to occur in a Ti-10V-2Fe-3Al alloy by Duerig et al. [2] and in Ti-Mo-Al alloys by Sasano and Suzuki [3]. In the late 1990s and early 2000s, superelasticity at room temperature has been confirmed in some Ti-Mo-based alloys [4, 5]. Since then, a great deal of research has been devoted to the development of biomedical shape memory alloys because of the concern over the high Ni concentration of practical Ti-Ni shape memory alloys. In the 2000s, Ti-Nb alloys have attracted attention due to their excellent biocompatibility. The Nb content dependence of the crystal structure and martensitic transformation temperature of Ti-Nb binary alloys have been clarified [6, 7]. And the effect of ternary alloying elements, such as Sn [8, 9], Ta [10], Zr [11], and Mo [12], on the martensitic transformation behavior and superelastic properties of Ti-Nb-based alloys has been

Ni-free Ti-based Shape Memory Alloys
https://doi.org/10.1016/B978-0-12-809401-3.00002-1

investigated. Thanks to extensive research efforts in the past decade, many Ti-Nb-based shape memory alloys have been developed, e.g., Ti-Nb-Sn [8, 9, 13], Ti-Nb-Al [14, 15], Ti-Nb-(Ga, Ge) [16], Ti-Nb-Ta [10, 17, 18], Ti-Nb-Zr [11, 19–23], Ti-Nb-Mo [12], Ti-Nb-O [24], Ti-Nb-N [25, 26], Ti-Nb-Pt [27], Ti-Nb-Pd [28], Ti-Nb-Ta-Zr [29, 30], Ti-Nb-Zr-Sn [31, 32], Ti-Nb-Mo-Sn [33, 34], Ti-Nb-Zr-Al [35], Ti-Nb-Zr-Ta-N [36], and Ti-Nb-Zr-Mo-Sn [37]. Very recently, Ti-Zr-based alloys have been developed with a greater emphasis on superelastic strain [38–41]. On the other hand, Ti-Ta shape memory alloys have attracted attention as potential high-temperature shape memory alloys [42–46]. In this chapter, an overview of the shape memory and superelastic properties of Ti-based alloys is provided. The superelastic properties of Ti-Nb binary alloys are first presented, and the effect of ternary alloying elements on superelastic properties of Ti-Nb alloys is discussed. Some important aspects of Ti-Mo-, Ti-Ta-, and Ti-Zr-based shape memory alloys are also summarized.

2.2 TI-NB BASED ALLOYS

2.2.1 Ti-Nb Binary Alloys

Shape memory effect and superelasticity have been reported to occur in many Ti-Nb-based alloys including Ti-Nb binary alloys. Shape memory effect and superelasticity are associated with the reversible martensitic transformation between the bcc β phase and the orthorhombic α'' martensite phase. Fig. 2.1 shows the dependence of martensitic transformation start temperature (M_s) on Nb content in the binary Ti-Nb alloys [47]. M_s decreases almost linearly with increasing Nb content and reaches room temperature when the Nb content is about 25 at.%. M_s decreases with a slope of −40 K/at.% Nb for Ti-(20–28)Nb alloys.

Fig. 2.2 shows stress-strain curves obtained by loading-unloading tensile tests at room temperature for Ti-Nb binary alloys subjected to the solution treatment [47]. An arrow with a broken line represents shape recovery upon heating up to 500 K. As expected from Fig. 2.1, superelasticity is observed at room temperature in the alloys whose Nb content is higher than 25 at.%, i.e., Ti-26Nb and Ti-27Nb, whereas shape memory effect is observed in the alloys with lower Nb contents as shown in Fig. 2.2. It is noted that the apparent yield stress, which corresponds to the stress for rearrangement of martensite variants, decreases with increasing Nb content from 20 to 25 at.%. The apparent yield stress for the Ti-26Nb and Ti-27Nb alloys corresponds to the critical stress for inducing martensitic transformation.

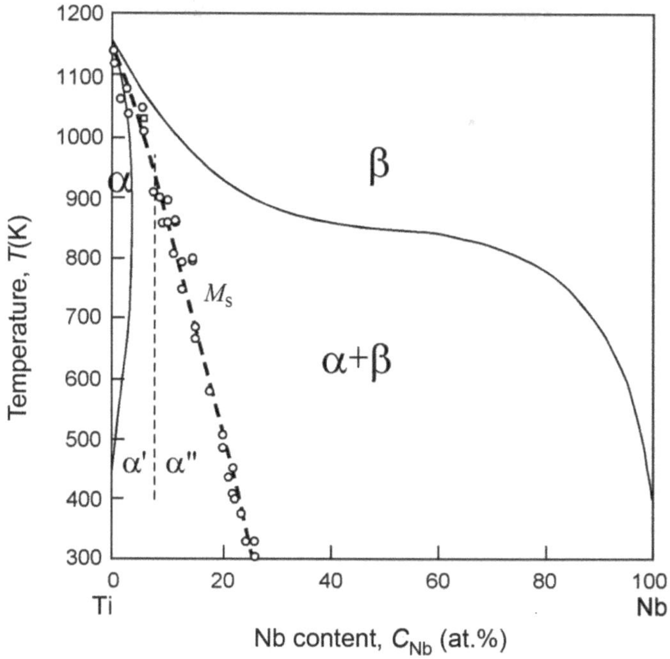

Fig. 2.1 Nb content dependence of martensitic transformation start temperature in Ti-Nb binary alloys.

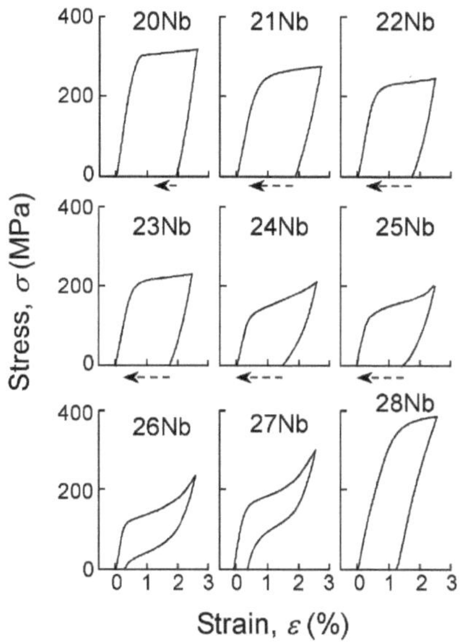

Fig. 2.2 Stress-strain curves obtained by loading-unloading tensile tests for Ti-Nb binary alloys.

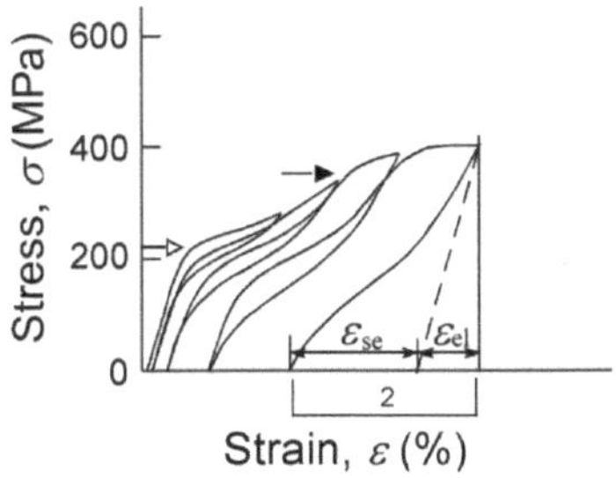

Fig. 2.3 Stress-strain curves obtained by strain increment cyclic loading and unloading tensile test of a Ti-27Nb alloy.

The transformation temperature decreases as the Nb content increases; thus, the apparent yield stress of the Ti-27Nb alloy is higher than that of the Ti-26Nb alloy. The Ti-28Nb alloy exhibits neither shape memory effect nor superelasticity because the critical stress for inducing martensitic transformation is higher than the critical stress for plastic deformation of β phase.

The superelastic properties can be evaluated by cyclic loading-unloading tensile tests. Fig. 2.3 shows cyclic stress-strain curves of the solution-treated Ti-27Nb alloy [47]. It is seen that the maximum superelastic transformation strain ε_{se} is only 1.2% in the Ti-27Nb alloy. This small superelastic transformation strain is mainly due to the intrinsic small lattice distortion caused by stress-induced martensitic transformation of the Ti-27Nb alloy. For the Ti-Nb binary alloys, the transformation strain exhibits a strong composition dependence and orientation dependence as shown in Fig. 2.4 [7]. For the β

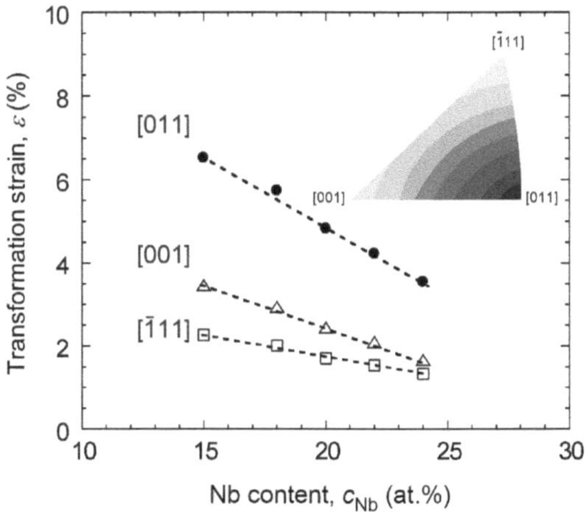

Fig. 2.4 Nb content dependence of the transformation strain for Ti-Nb binary alloys.

to α'' transformation, the largest transformation strain is obtained along the [011] direction of the β phase. The transformation strain decreases linearly with the Nb content. Extrapolation of linear regression plot reveals an estimated transformation strain of 2.6% along the [011] direction at 27 at.% Nb, implying that the largest superelastic transformation strain obtainable is 2.6% even in the single crystal for the Ti-27Nb alloy. This value is far below that of the B2-B19′ transformation in Ti-Ni alloys. This small transformation strain is one of the drawbacks of the Ti-27Nb alloy. The transformation strain can be increased by reducing the Nb content; however, the decrease in the Nb content raises the martensitic transformation temperature as shown in Fig. 2.1, and thus, superelasticity cannot be obtained at room temperature. It is also noted that the low critical stress for the plastic deformation of the Ti-27Nb alloy is another drawback. The critical stress for the plastic deformation can be raised through microstructure control such as aging and low-temperature annealing. However, the addition of alloying elements is indispensable in order to increase the transformation strain by changing the lattice constants of the β and α'' phases.

2.2.2 Ti-Nb-Zr Alloys

In order to improve superelastic properties of Ti-Nb-based alloys, the addition of ternary elements has been considered. Among many alloying elements, it has been reported that Zr is the most effective alloying element for increasing superelastic transformation strain. The composition dependence of superelastic properties and crystal structures of β and α'' phases in the Ti-Nb-Zr ternary system have been extensively investigated. Fig. 2.5 shows the compositions in a Ti-Nb-Zr ternary diagram exhibiting shape memory effect and superelasticity where the compositions showing shape memory effect and superelasticity are marked by open circles and solid circles, respectively [48]. It is seen that the shape memory effect and superelasticity are observed in a wide composition range of Ti-Nb-Zr ternary alloys. Some examples of the stress-strain curves of Ti-Nb-Zr alloys obtained by loading-unloading tensile tests at room temperature are shown in Fig. 2.6 [48]. For the Ti-Nb-12Zr alloys, the shape memory effect was observed when the Nb content is 15 and 16 at.%, and superelastic recovery was observed when the Nb content is 17 and 18 at.%. Comparing with the results of binary Ti-Nb alloys (Fig. 2.2), it is clear that the superelastic recovery is observed in the alloys with a lower Nb content, implying that Zr reduces the martensitic transformation temperature. The Nb content

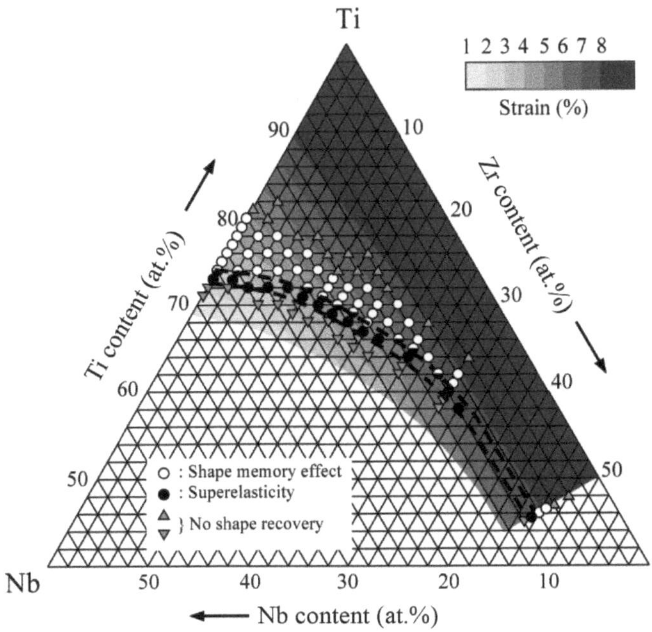

Fig. 2.5 Composition dependence of the transformation strain along the [011] direction and shape memory properties in Ti-Nb-Zr alloys.

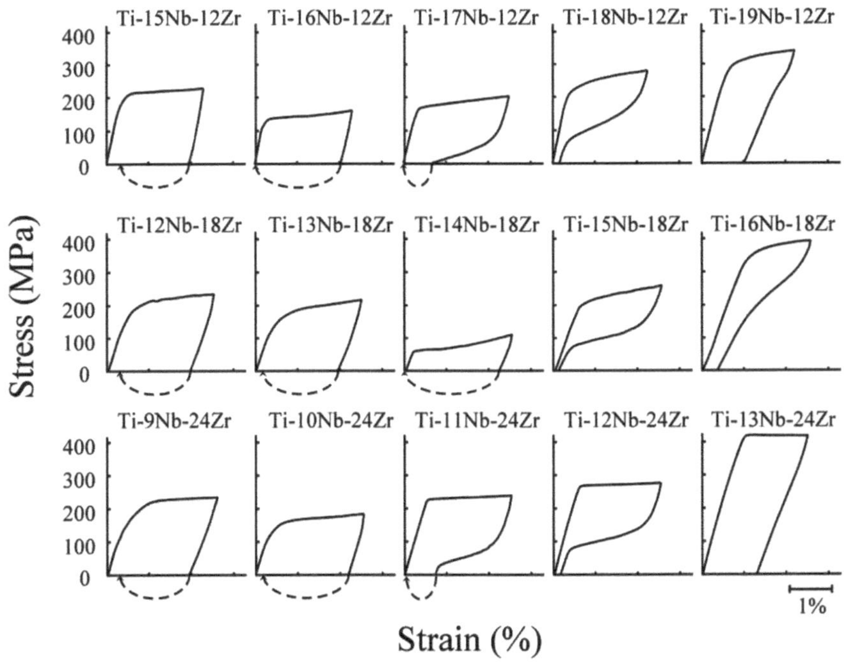

Fig. 2.6 Stress-strain curves obtained by loading-unloading tensile tests for Ti-Nb-Zr alloys.

exhibiting superelastic recovery further decreases with increasing Zr content, i.e., 15 and 12 at.% when the Zr contents are 18 and 24 at.%, respectively. It has been confirmed that the addition of 1 at.% Zr to a Ti-22Nb alloy reduces the martensitic transformation start temperature by 35 K.

Fig. 2.5 also shows that the composition dependence of the transformation strain along the [011] direction, which is the maximum transformation strain for each composition, is also shown by means of contour lines. It is seen that the Zr content has a weak effect on the transformation strain particularly in the alloys with a lower Nb content. It is noted that the transformation strain along the [011] direction increases as the Zr content increases when compared with the composition range showing superelasticity. The Ti-27Nb alloy exhibits a transformation strain of 2.6%, but it increases to 5.1% for the Ti-15Nb-18Zr alloy. The transformation strain increases to more than 7% when the Zr content is higher than 30 at.%. The positive effect on the transformation strain of the Zr addition has been confirmed by the cyclic tensile tests.

2.2.3 Ti-Nb-Ta Alloys

Ta has been considered as an attractive alloying element for improving shape memory and superelastic properties owing to its excellent biocompatibility. The effect of Ta addition on the superelastic properties and crystal structures of β to α'' phases has been extensively investigated. Fig. 2.7 shows the compositions exhibiting shape memory effect and superelasticity in a Ti-Nb-Ta composition map [48]. Similar to the Ti-Nb-Zr system, the shape memory effect and superelasticity are observed in a rather wide composition range of Ti-Nb-Ta ternary alloys. However, it is noted from Fig. 2.8 [48] that the superelastic properties of Ti-Nb-Ta alloys tend not to be as good as Ti-Nb-Zr alloys. As shown in Fig. 2.8, Ti-19Nb-10Ta and Ti-13Nb-20Ta alloys exhibited superelastic shape recovery, but the shape recovery was not completed. Furthermore, the superelasticity was rarely observed in the alloys containing more than 20 at.% Ta. It is also clear from Fig. 2.7 that Ta decreases the martensitic transformation temperature because superelasticity was observed in alloys with a lower Nb content by the addition of Ta. It has been reported that the addition of Ta reduces the transformation start temperature with a slope of 30 K per 1 at.% [49]. Relatively poor superelastic properties of Ti-Nb-Ta alloys are due to a small transformation strain. As seen in Fig. 2.7, the increase of the Ta content in the Ti-Nb-Ta ternary alloys decreases the transformation strain in the composition range revealing superelasticity.

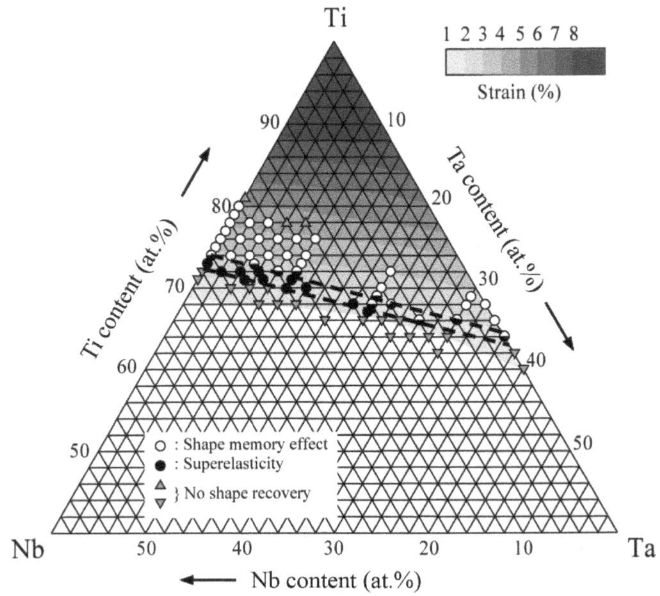

Fig. 2.7 Composition dependence of the transformation strain along the [011] direction and shape memory properties in Ti-Nb-Ta alloys.

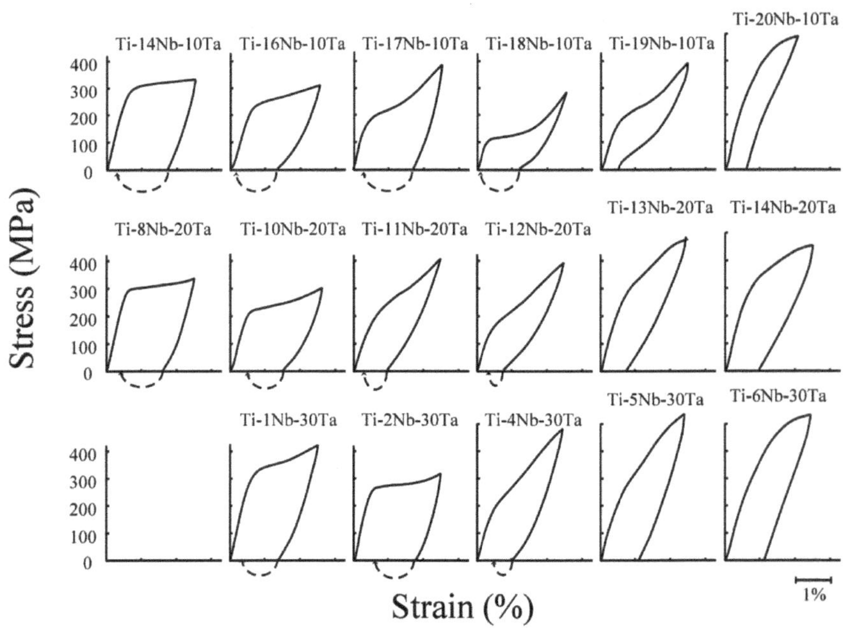

Fig. 2.8 Stress-strain curves obtained by loading-unloading tensile tests for Ti-Nb-Ta alloys.

2.2.4 Ti-Nb-Mo Alloys

Mo is a strong β stabilizer and an attractive alloying element for Ti alloys. Fig. 2.9 shows stress-strain curves obtained by loading-unloading tensile tests at room temperature for Ti-Nb-Mo alloys annealed at 973 K [12]. It is seen that the composition showing superelasticity is shifted toward lower Nb content with increasing Mo content. Ti-(27, 28)Nb, Ti-(23, 24)Nb-1Mo, Ti-(19–22)Nb-2Mo, and Ti-(15–18)Nb-3Mo alloys exhibit superelastic recovery among Ti-Nb-Mo alloys with Mo contents of 0, 1, 2, and 3 at.%, respectively, implying that the impact of Mo on the transformation temperature is 3–4 times stronger than Nb. This was confirmed by differential scanning calorimetry measurements as shown in Fig. 2.10 [12]. It is seen that the reverse martensitic transformation finish temperature (A_f) of the Ti-22Nb alloy decreases by about 100 K by the addition of 1 at.% Mo. On the other hand, the A_f decreases by 25–30 K per 1 at.% increase in Nb content in the Ti-(21–25)Nb, Ti-(17–22)Nb-1Mo, and Ti-(14–16) Nb-2Mo alloys.

The addition of Mo is beneficial in increasing the critical stress for slip deformation due to the solid-solution strengthening effect. The maximum superelastic recovery strain increases by increasing Mo content due to the increase in the critical stress for slip and transformation strain. The increase in the Mo content accelerates the formation of the ω phase. Fig. 2.11 shows dark-field images exhibiting the ω phase in the Ti-27Nb, Ti-24Nb-1Mo, Ti-21Nb-2Mo, Ti-18Nb-3Mo, and Ti-13Nb-4Mo alloys, which exhibit the most stable superelasticity among Ti-Nb-Mo alloys with Mo contents

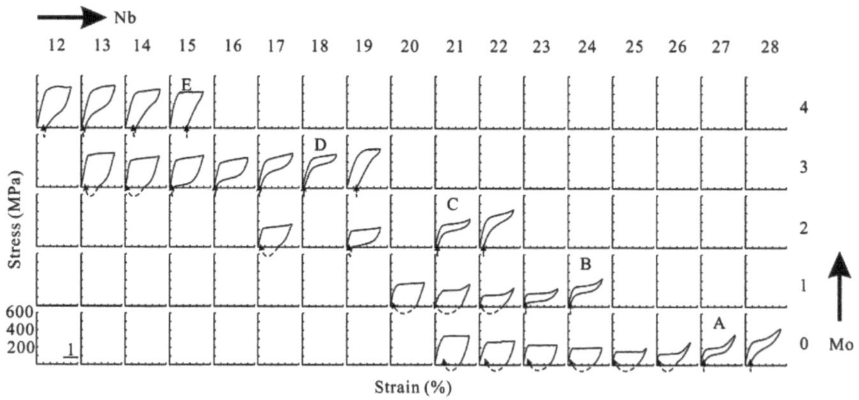

Fig. 2.9 Stress-strain curves obtained by loading-unloading tensile tests for Ti-Nb-Mo alloys.

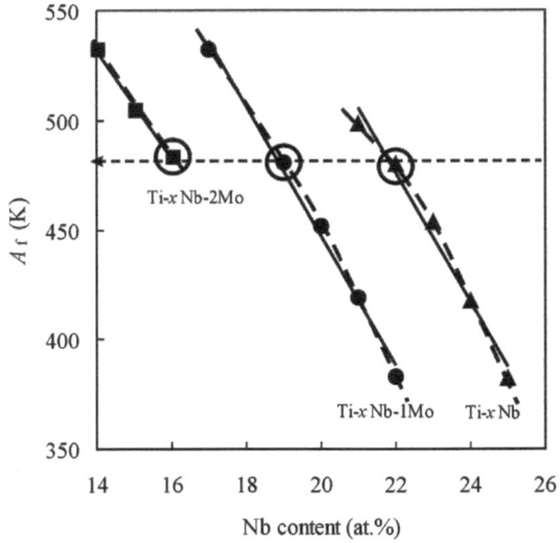

Fig. 2.10 Nb content dependence of reverse martensitic transformation finish temperature for binary, 1Mo-added and 2Mo-added alloys.

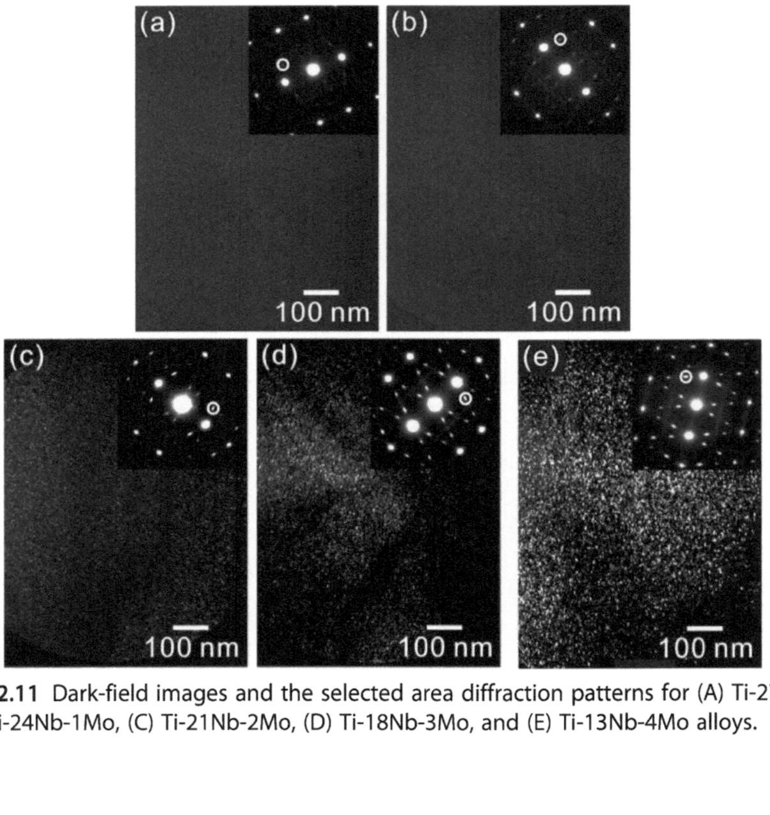

Fig. 2.11 Dark-field images and the selected area diffraction patterns for (A) Ti-27Nb, (B) Ti-24Nb-1Mo, (C) Ti-21Nb-2Mo, (D) Ti-18Nb-3Mo, and (E) Ti-13Nb-4Mo alloys.

of 0, 1, 2, 3, and 4 at.%, respectively [50]. It is clearly seen that the volume fraction of the ω phase increases with increasing Mo content or with decreasing Nb content. This is due to the decrease in total amount of β-stabilizing alloying elements, i.e., the decrease in electron-to-atom (*e*/*a*) ratio of the alloy. The increase in the volume fraction of athermal ω phase increases the stress for inducing martensitic transformation, causing a wide stress hysteresis as can be seen in Fig. 2.9.

The stress hysteresis can be reduced by the suppression of athermal ω phase. The addition of Al and Sn is very effective for suppressing the formation of athermal ω phase. Fig. 2.12 shows the effect of Sn addition on the stress hysteresis and superelastic properties of a Ti-15Nb-3Mo alloy [34]. It is clearly seen that stress hysteresis decreases monotonically with increasing Sn content. It is also noted that the critical stress for inducing martensitic transformation, indicated by a single-headed arrow, does not show a monotonic change. The critical stress for inducing martensitic transformation decreases by the addition of Sn up to 1 at.% and then increases gradually with further addition of Sn. This result indicates that the martensitic transformation start temperature increases by the addition of Sn up to 1 at.%. This tendency is opposite to the fact that Sn decreases the martensitic transformation start temperature of Ti-Nb-based alloys. This anomalous Sn content dependence of the stress-induced martensitic transformation is explained by a dual role of Sn, first in decreasing the martensitic transformation temperature and second in suppressing the formation of athermal ω phase. It is considered that the effect of Sn on the suppression of athermal ω phase is dominant when the Sn content is up to 1 at.%. The amount of athermal ω phase decreases with increasing Sn content, causing the effect of Sn on the suppression of athermal ω phase to become weaker. Therefore, the effect of Sn on the decrease in the martensitic transformation becomes dominant, and then, the critical stress for inducing martensitic transformation increases with increasing Sn content

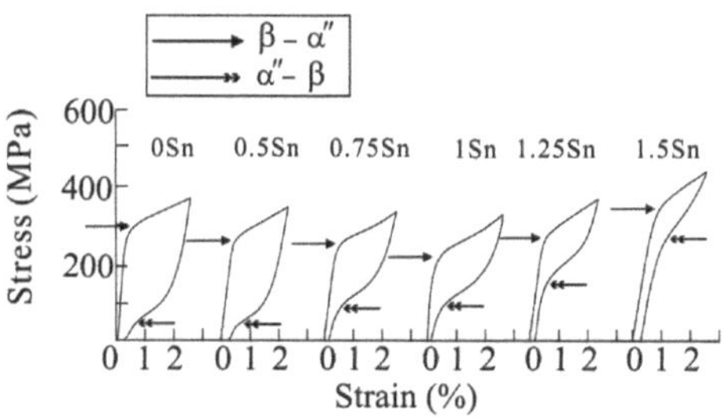

Fig. 2.12 Effect of Sn content on superelastic behavior of Ti-15Nb-3Mo-(0–1.5)Sn alloys.

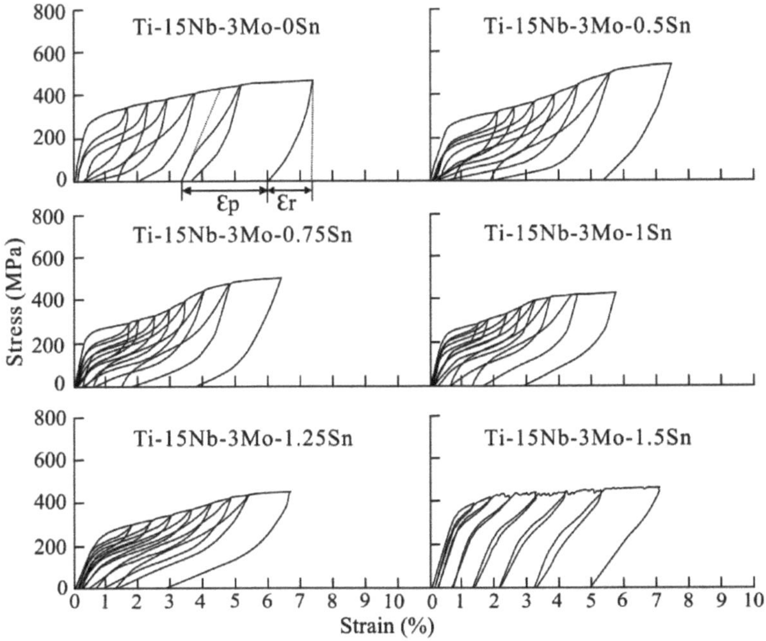

Fig. 2.13 Stress-strain curves obtained by strain increment cyclic loading and unloading tensile tests of Ti-15Nb-3Mo-(0–1.5)Sn alloys.

from 1 at.%. Different from the critical stress for inducing martensitic transformation, the stress where the reverse transformation finishes increases monotonically with increasing Sn content as indicated by a double-head arrow. This suggests that the effect of the athermal ω phase on the reverse transformation is small because the ω phase transforms to α″ phase along with the β phase upon loading. The decrease in the stress hysteresis leads to an improvement of superelastic recovery strain as shown in Fig. 2.13 [34]. Ti-15Nb-3Mo exhibits the maximum recovery strain of 2.1%, while the magnitude of the maximum recovery strain increases to 3.7% by the addition of 1.25 at.% Sn.

2.3 TI-MO BASED ALLOYS

The martensitic transformation from β to α″ has been confirmed in Ti-Mo binary alloys [51]; however, they exhibited neither a clear shape memory effect nor superelasticity. Shape memory effect and superelasticity have been reported in some Ti-Mo-based ternary and multinary alloys such as Ti-Mo-Al [3], Ti-Mo-Sn [52], Ti-Mo-Ga [53], Ti-(8–10)Mo-3Al-2V-4Nb (wt%) [54], and Ti-4Mo-5Al-2Sn-4Zr-2Cr-1Fe (wt%) (β-Cez) alloys [5]. Shape memory effect in Ti-Mo-based alloys was first reported by Sasano and Suzuki [3] in 1985. They investigated the effect of Mo content on the microstructure and shape memory properties in Ti-(10–15)Mo-3Al (wt%) alloys. Ti-(10, 11)Mo-3Al (wt%) alloys revealed a full martensite structure. Ti-(12, 13) Mo-3Al (wt%) alloys showed a partial martensite structure, and a Ti-14Mo-3Al (wt%) alloy shows a single β phase; it is therefore obvious that Mo decreases the martensitic transformation start temperature. The Ti-14Mo-3Al (wt%) alloy exhibited superelastic recovery upon unloading after an initial strain of 3% at 150 K, and a small remained strain was fully recovered at 300 K. The reverse transformation finish temperature (A_f) increases with decreasing Mo content: A_f increases by about 75 K with a decrease of 1 wt% Mo. The Ti-(12–15)Mo-3Al (wt%) alloys exhibited perfect shape recovery up to about 3% strain. The stress-induced martensitic transformation from β to α″ was confirmed in a Ti-15Mo-3Al (wt%) alloy. And they showed that the increase in the Mo content decreased the lattice distortion needed to transform from β to α″. They also investigated the effect of Al on the shape memory properties of Ti-14Mo and showed that Al decreased A_f and increased the shape recovery strain.

Lei et al. [4] have investigated the effect of alloying elements such as Cr, V, and Nb on superelastic properties of Ti-Mo-Al alloys aiming at orthodontic arch wires. Among the investigated alloys, a Ti-11Mo-3Al-2V-4Nb (wt%) alloy exhibited the best superelastic properties with a maximum recovery strain of 3.5% although the applied strain was not recovered completely. They also reported that aging at temperatures in the range of 473–673 K decreased superelastic recovery strain and ductility due to the formation of thermal ω phase. Zhou et al. [54] have investigated the superelastic properties of Ti-(8–11)Mo-3Al-2V-4Nb (wt%) alloys and claimed that the superelastic recovery is highly sensitive to the Mo content. A Ti-9Mo-3Al-2V-4Nb alloy exhibited most excellent superelasticity with a recovery strain of 2.5% after tensile testing to 4% strain.

Shape memory effect and superelasticity at room temperature have been confirmed in Ti-Mo-Ga alloys [53, 55]. Fig. 2.14 shows a result of bending and heating test of a Ti-6Mo-3Ga alloy annealed at 1073 K for 120 s [55]. It is seen that the specimen bent at room temperature perfectly recovered to the original straight shape by heating up to 500 K. The Ti-6Mo-3Ga alloy exhibits a perfect shape memory effect within 3.5% strain by the bending test. The shape memory effect was also observed in the specimen deformed by tensile tests as shown in Fig. 2.15 where a broken line with an arrow

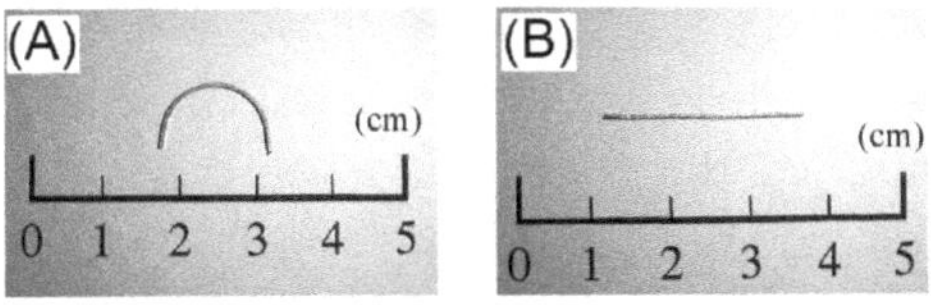

Fig. 2.14 A result of bending test for a Ti-6Mo-3Ga alloy: (A) after deformation and (B) after heating up to 500 K.

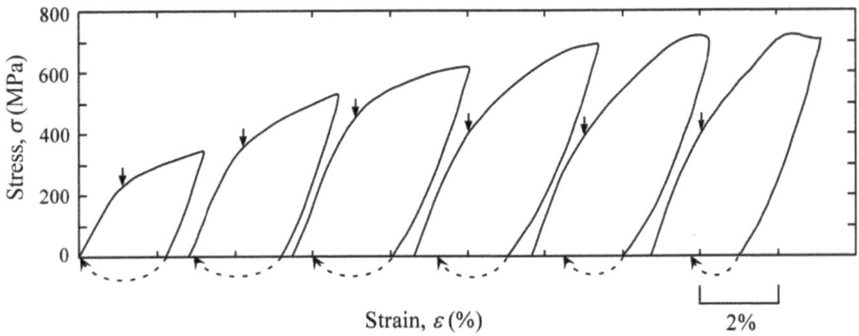

Fig. 2.15 Stress-strain curves obtained by cyclic loading and unloading tensile tests of a Ti-6Mo-3Ga alloy.

indicates the shape recovery upon heating at 500 K [55]. The tensile strain was fully recovered at the first cycle, but the shape recovery became incomplete as the number of cycle increased. It is also noted that the critical stress for inducing martensitic transformation, indicated by an arrow, increases with increasing number of cycles. The increase in the critical stress for inducing martensitic transformation is due to the formation of thermal ω phase formed upon heating at 500 K. The Ti-6Mo-3Ga alloy was found to be very susceptible to ω-phase embrittlement. As shown in Fig. 2.16 [55], the hardness increased significantly by aging at 473–773 K although the aging time was only 60 s. The formation of thermal ω phase upon aging at temperatures up to about 723 K has also been also reported in Ti-Mo-based alloys. Fig. 2.16 also shows that the addition of Nb is effective to suppress the formation of thermal ω phase. The addition of Nb decreased the critical stress for inducing martensitic transformation and increased the shape recovery strain. Superelasticity at room temperature was observed in a Ti-7Mo-4Ga alloy with a recovery strain of 4% [53].

Maeshima and Nishida have investigated the effect of Ag, Sn, and Sc content on the shape memory properties of Ti-Mo-(Ag, Sn, and Sc) alloys

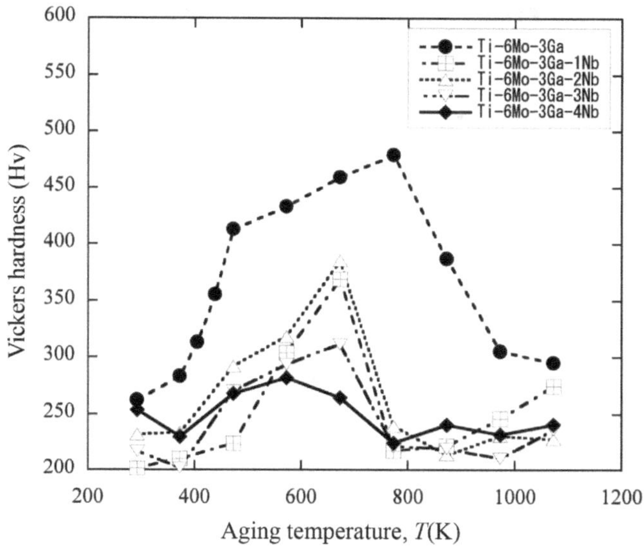

Fig. 2.16 Vickers hardness of the Ti-6Mo-3Ga-(0–4)Nb alloys subjected to aging treatment at temperatures between 373 and 1073 K for 60 s.

[52, 56, 57]. Ti-6Mo-5Sc, Ti-5Mo-(2–5)Ag, and Ti-5Mo-(1–3)Sn alloys exhibited shape memory effect with a recovery strain of 3%–5%. They showed that the increase in the Sn content decreased the martensitic transformation temperature of Ti-5Mo-(0–6)Sn alloys and superelasticity was observed in a Ti-5Mo-5Sn alloy with a recovery strain of 3%. The superelastic recovery strain of the Ti-5Mo-5Sn alloy increased to 3.5% by a short-time aging at 873 K for 300 s. Sutou et al. [58] investigated the mechanical properties of a cold-drawn Ti-6Mo-4Sn wire. They show that a maximum recovery strain of about 4% including elastic strain was obtained in the wires annealed in the temperature range of 1073–1273 K.

Stress-induced martensitic transformation from β to α″ and superelastic recovery were reported to occur in a β-Cez alloy [5]. Grosdidier and Philippe [5] reported that superelastic recovery can be obtained in a β-Cez alloy after solution treatment above β-transus temperature. A superelastic strain of 2.5% was obtained upon unloading after an initial strain of about 3% for the specimen solution treated at 1193 K. The recovery strain decreased with decreasing annealing temperature, and no shape recovery was observed in the specimen annealed at 1073 K. This is due to the fact that the β phase became stable by the precipitation of α phase, causing the stress-induced martensitic transformation not to occur.

2.4 TI-TA BASED ALLOYS

Shape memory alloys are very useful for actuators due to their high recovery force, large recovery strain, and high work output. However, commercially available Ti-Ni-based alloys are limited in their application to temperature below 373 K because of their low transformation temperatures. Therefore, the development of high-temperature shape memory alloys that can operate in a hot environment has been actively pursued for use in the field of power generation, automotive, and aerospace. It has been well established that the addition of precious alloying elements such as Pd, Pt, and Au increases the transformation temperature. And the addition of Zr and Hf in replacement of Ti leads to increase in the transformation temperature of Ti-Ni alloys. However, high cost of raw materials and inherent poor cold workability of Ti-Ni-based high-temperature shape memory alloys have limited their practical applications to temperature below 373 K.

Ti-Ta alloys have attracted attention as high-temperature shape memory alloys. Fig. 2.17 shows a composition dependence of martensitic transformation start temperature determined from strain-temperature curves upon thermal cycling for Ti-Ta and Ti-Nb solution-treated specimens [42]. It is clear that Ta is a weaker β-stabilizer when compared with Nb. The martensitic transformation temperature decreases by about 30 K per 1 at.% Ta for the Ti-Ta alloys, while it decreased by about 40 K per 1 at.% Nb for the Ti-Nb alloys. The alloys containing less than 35 at.% Ta for the Ti-Ta alloys and 23 at.% Nb for the Ti-Nb alloys are possible candidate for high-temperature shape memory alloys. The stability of shape memory effect is particularly important to ensure practical applications. The thermal stability of shape memory effect can be evaluated by thermal cycling tests under a constant stress. Fig. 2.18A shows strain-temperature curves for Ti-22Nb solution-treated specimens thermally cycled between 173 and 513 K under a constant stress of 50 MPa [42]. The shape memory effect was confirmed during the first thermal cycle with M_s of 420 K and A_f of 475 K. During the second thermal cycle, the Ti-22Nb alloy did not exhibit shape memory effect. The disappearance of shape memory effect is due to aging effects upon thermal cycling. It was confirmed by TEM observation that the ω phase is formed in the β phase during thermal cycling and it suppressed the martensitic transformation. The increase in the maximum temperature to 573 K in the thermal cycle range completely suppressed the martensitic transformation even during the first thermal cycle as shown in Fig. 2.18B due to the acceleration of ω-phase precipitation [42].

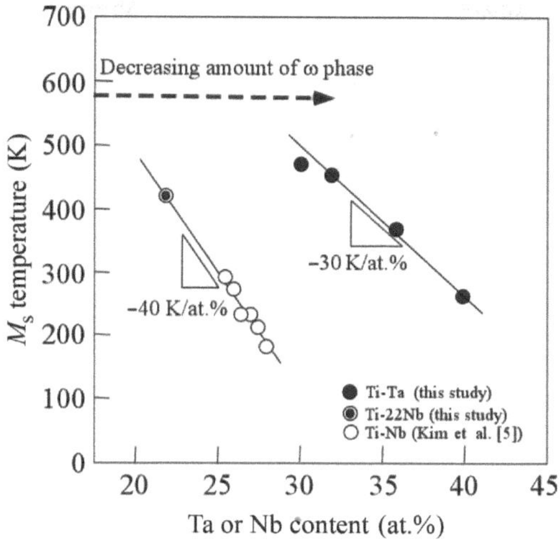

Fig. 2.17 Martensitic transformation start temperature for Ti-Nb and Ti-Ta alloys. The *arrow* indicates the direction of decreasing ω-phase stability during aging at 573 K for 3.6 ks.

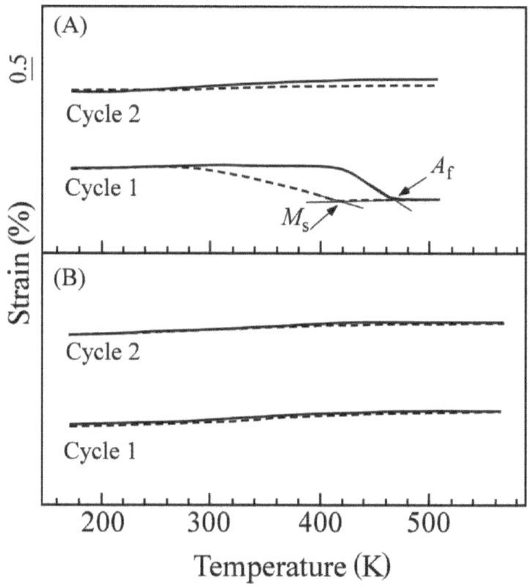

Fig. 2.18 Thermal cycle tests under a constant stress of 50 MPa for Ti-22Nb solution-treated specimens, which were thermally cycled (A) between 173 and 513 K and (B) between 173 and 573 K, respectively.

Fig. 2.19 shows strain-temperature curves for Ti-32Ta solution-treated specimens measured by thermal cycling in the temperature range of 173–513 and 173–573 K under a constant stress of 50 MPa [42]. The Ti-32Ta alloy exhibited high-temperature shape memory effect at the first cycle; M_s and A_f were measured to be 440 and 490 K, respectively. Different from the

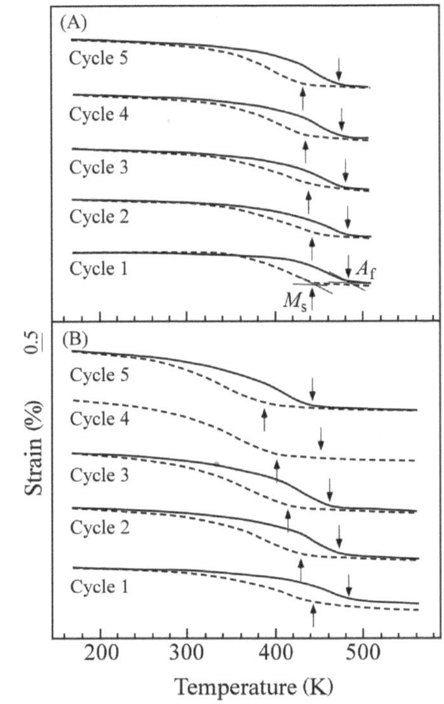

Fig. 2.19 Thermal cycle tests under a constant stress of 50 MPa for Ti-32Ta solution-treated specimens, which were thermally cycled (A) between 173 and 513 K and (B) between 173 and 573 K, respectively.

results of the Ti-22Nb alloy, the shape memory effect was observed during successive cycles although the transformation temperatures decreased with increasing number of cycles, indicating that shape memory effect of Ti-Ta alloys is more stable when compared with Ti-Nb alloys with similar transformation temperatures. It is also noted that the decrease in the transformation temperature became more significant by increasing the maximum temperature in thermal cycling range to 573 K. It was also confirmed that the decrease in the transformation temperatures is due to the formation of the ω phase during thermal cycling.

In order to improve the thermal stability of Ti-Ta alloys, effects of ternary alloying element on the martensitic transformation, shape memory properties, and microstructural stability have been symmetrically investigated. Fig. 2.20 shows the effect of alloying element on M_s of Ti-30Ta-X (X=V, Cr, Fe, Zr, Hf, Mo, Sn, and Al) alloys [43]. M_s decreases almost linearly for all the alloying elements. It has been suggested that the number of

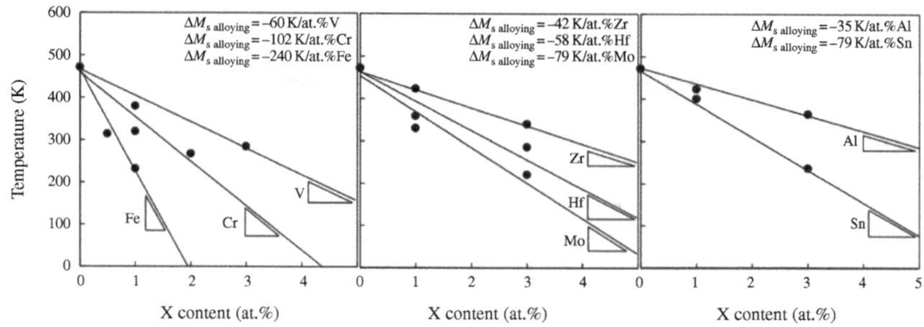

Fig. 2.20 Martensitic transformation start temperature as a function of alloying content for Ti-30Ta-X solution-treated alloys. The slope corresponds to the amount of decrease in M_s for 1 at.% X element, and for this study, it is termed as $\Delta M_{s\ alloying}$.

valence electrons and the atomic size of the alloying element are main factors decreasing M_s of Ti-Ta-X alloys: alloying elements with a high number of valence electrons and a small atomic size strongly decrease M_s. Alloying elements also affect aging effect and thermal stability. As shown in Fig. 2.21, for all alloys, M_s decreased after aging at 573 K for 3.6 ks [43]. After aging, M_s of the Ti-30Ta binary alloy decreased more than 100 K when compared with the solution-treated condition due to the formation of the ω phase. The addition of V, Cr, Fe, Zr, Hf, and Mo did not suppress the effect of aging on M_s since they did not suppress the formation of the ω phase during aging treatment. On the other hand, the addition of Al and Sn is very effective for the improvement of thermal stability since the addition of Al and Sn

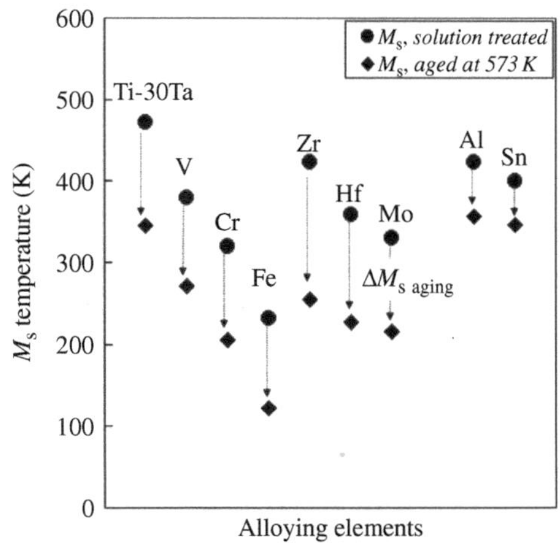

Fig. 2.21 Martensitic transformation start temperature of Ti-30Ta-1X solution-treated specimens and aged (573 K for 3.6 ks) specimens.

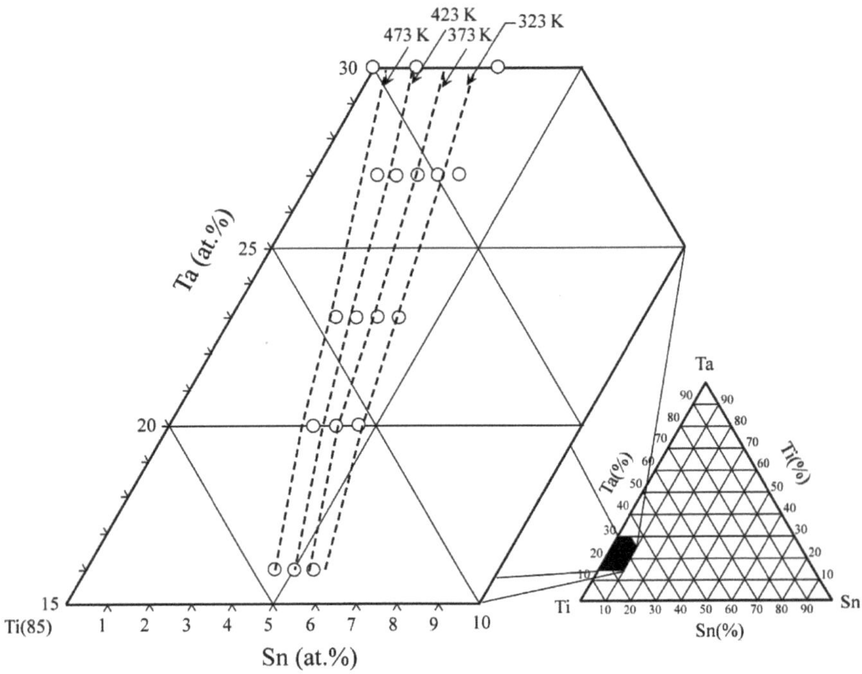

Fig. 2.22 Composition dependence of martensitic transformation start temperature in Ti-Ta-Sn alloys.

suppresses the formation of the ω phase in Ti-Ta alloys similar to Ti-Nb and Ti-Mo alloys.

The effects of composition on the martensitic transformation temperature, crystal structure, and thermal stability in Ti-Ta-Sn and Ti-Ta-Al ternary alloys were considerably investigated [44–46]. Fig. 2.22 shows the composition dependence of M_s in Ti-Ta-Sn alloys. M_s decreased by about 100 K per 1 at.% Sn and by about 25 K per 1 at.% Ta in the Ti-(16–27)Ta-(1.5–5.5)Sn alloys [44]. This implies that the effect of Sn on the decrease of M_s is four times stronger than the effect of Ta in the Ti-Ta-Sn alloys and M_s of Ti-Ta-Sn alloys can be tailored by adjusting Ta and Sn contents. Ti-16Ta-(4.5, 5)Sn, Ti-20Ta-3.5Sn, Ta-23Ta-(2.5, 3)Sn, Ti-27Ta-(1.5, 2)Sn, and Ti-30Ta-1Sn alloys exhibit a high M_s above 373 K. Thermal stability, i.e., stability against the formation of ω phase, is dependent on both Ta and Sn contents. Fig. 2.23 shows M_s of specimen solution treated (solid symbols) and aged at 573 K for 3.6 ks (open symbols) [44]. M_s decreases after aging for all compositions; however, the amount of the decrease in M_s after aging became smaller with increasing amount of Sn and Ta. It is noted that

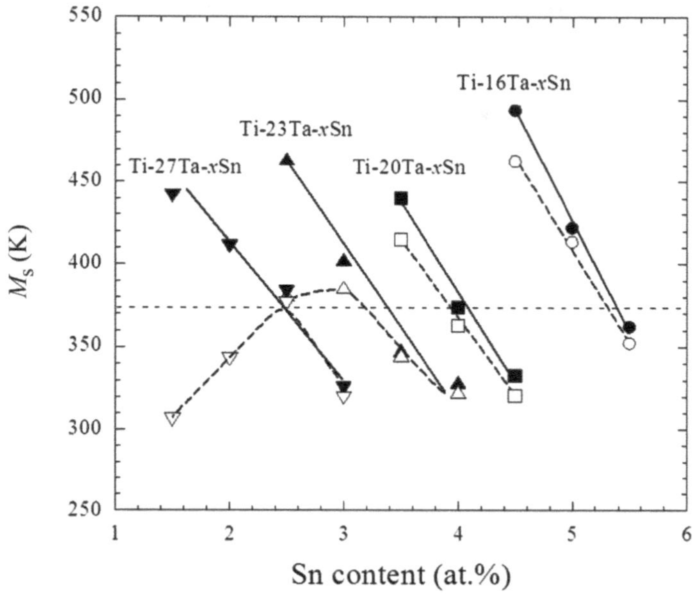

Fig. 2.23 Change of martensitic transformation start temperature before (*closed symbols* and *solid lines*) and after aging at 573 K for 3.6 ks (*open symbols* and *dashed lines*) in Ti-Ta-Sn alloys.

the amount of Sn content needed to suppress the formation of ω phase increased with decreasing Ta content. Ti-16Ta-4.5Sn, Ti-16Ta-5Sn, Ti-20Ta-3.5Sn, Ti-23Ta-3Sn, and Ti-27Ta-2.5Sn alloys show a higher M_s above 373 K after aging.

Fig. 2.24 shows the strain-temperature curves at 100 MPa of (a) Ti-32Ta, (b) Ti-27Ta-1.5Sn, and (c) Ti-20Ta-3.5Sn solution-treated specimens obtained by thermal cycling in the temperature range of 173–513 K under a stress of 100 MPa [44]. It is noted that all alloys exhibit similar M_s (430–450 K) at the first cycle; however, the decrease in M_s upon thermal cycling shows composition dependence. As the number of cycle increased, M_s of the Ti-32Ta and Ti-27Ta-1.5Sn decreased continuously while that of the Ti-20Ta-3.5Sn did not change up to the 20th cycle. Fig. 2.24 also reveals that the addition of Sn is beneficial for the increase in the shape recovery strain. A larger shape recovery strain is due to the fact that the addition of Sn as a substitute of Ta with keeping a constant M_s increases the transformation strain. Thus, it is concluded that the addition of Sn in Ti-Ta alloys is very effective for not only improving the thermal stability of shape memory effect but also increasing the actuation stroke. It was also reported that

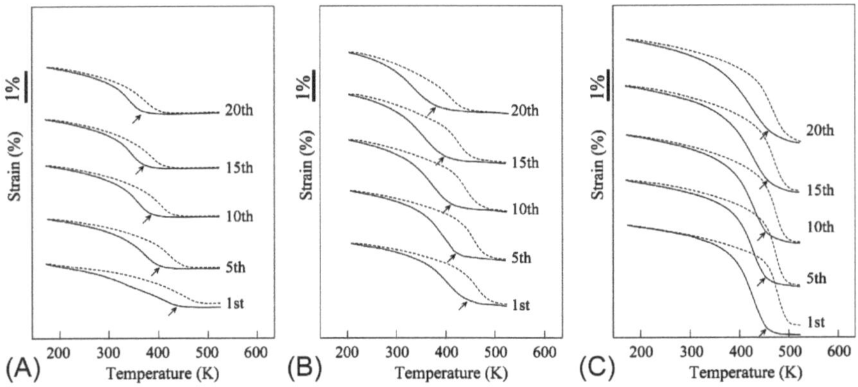

Fig. 2.24 Strain-temperature curves of (A) Ti-32Ta, (B) Ti-27Ta-1.5Sn, and (C) Ti-20Ta-3.5Sn alloys thermal cycled under the stress of 100 MPa up to 20 times. The *arrow* indicates martensitic transformation start temperature.

Ti-Ta-Sn alloys can be actuated by electric current heating. Al was also reported to be useful for improving thermal stability and shape memory effect of Ti-Ta alloys [45].

2.5 TI-ZR BASED ALLOYS

As mentioned in Section 2.2, many Ti-Nb-based alloys are reported to exhibit superelasticity at room temperature. However, the superelastic recovery strain in most of Ti-Nb-based superelastic alloys is limited in the range of 3%–4%, which is due to the small transformation strain at compositions showing superelasticity. It has been reported that a Ti-Zr system is promising candidate for superelastic alloys with a large recovery strain [39–41, 48]. The α/β transition temperature decreases with increasing Zr content and shows minimum value when the Zr content is 50 at.% in Ti-Zr binary alloys. It has also been reported that the martensitic transformation temperature decreases with increasing Zr content. Therefore, superelasticity is obtained in alloys with a smaller β-stabilizing alloying element such as Nb and Mo, resulting in a larger transformation strain.

Shape memory effect and superelasticity were confirmed in Ti-Zr-Nb-Sn alloys. Fig. 2.25 shows the effect of Sn content on superelastic properties of Ti-18Zr-(9.5–15)Nb-(0–4)Sn alloys [39]. It is noted that the Nb content was decreased with increasing Sn content in order to maintain the

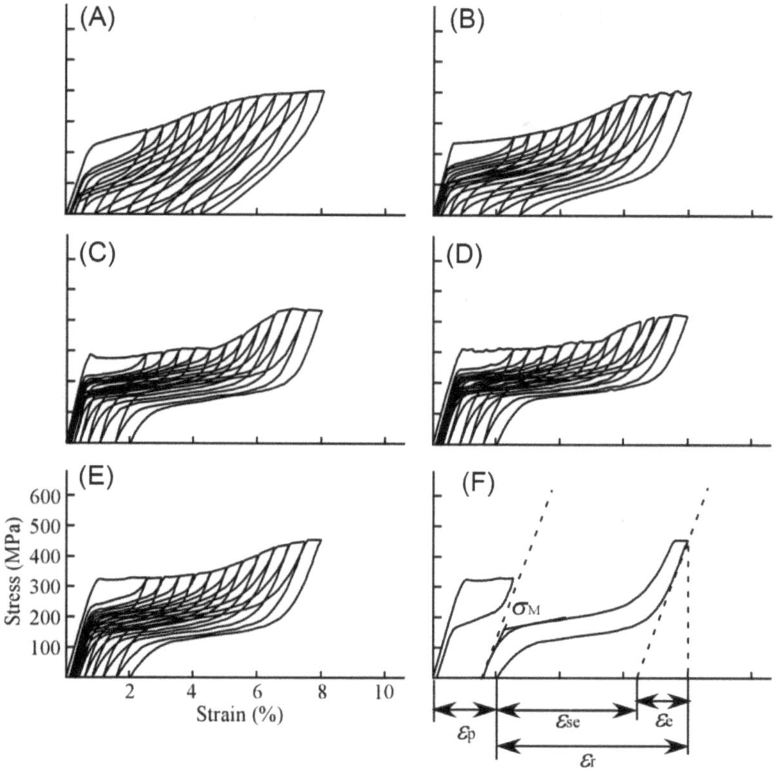

Fig. 2.25 Stress-strain curves obtained from strain increment cyclic tensile tests of Ti-18Zr-*x*Nb-*y*Sn alloys. (A) Ti-18Zr-15Nb, (B) Ti-18Zr-13.5Nb-1Sn, (C) Ti-18Zr-12.5Nb-2Sn, (D) Ti-18Zr-11Nb-3Sn, (E) Ti-18Zr-9.5Nb-4Sn, and (F) measurement scheme of superelastic properties.

martensitic transformation temperature to exhibit superelasticity at room temperature. The Ti-18Zr-15Nb alloy exhibits superelasticity with a superelastic strain (ε_{se}) of 2.9%. As Sn content increases, the superelastic strain increases up to about 5.0% and then decreases slightly. The increase in the superelastic strain by the increase of the Sn content is due to the combination effect of a large transformation strain and a desirable recrystallization texture. It was confirmed that the Ti-18Zr-15Nb and Sn-added alloys exhibit large transformation strains more than two times larger than that of a binary Ti-27Nb superelastic alloy. For example, the transformation strain along the [011] direction, due to the stress-induced martensitic transformation from the β phase to α″ martensite phase, was determined to be 5.3% in the Ti-18Zr-11Nb-3Sn. As shown in Fig. 2.26 [39], the

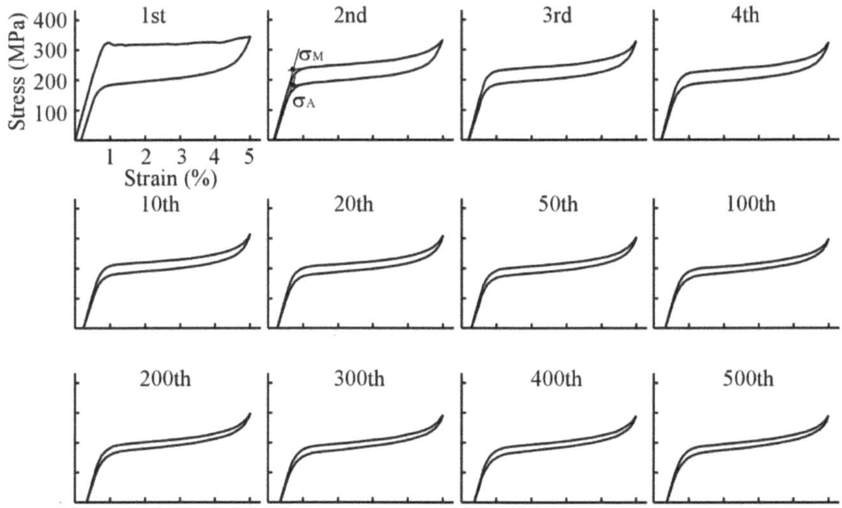

Fig. 2.26 Effect of cyclic deformation on the stress-strain curve of a Ti-18Zr-11Nb-3Sn alloy.

superelasticity of the Ti-18Zr-11Nb-3Sn alloy is found to be very stable during loading-unloading cycling up to 500 cycles. It is also noted that the Ti-18Zr-11Nb-3Sn alloy exhibits an extremely narrow stress hysteresis about 20 MPa. A narrow stress hysteresis is desirable for improving durability of superelastic devices because the energy loss is small during loading and unloading cycles. A large recovery strain of about 7%, which is comparable with the value obtained in Ti-Ni-based superelastic alloys, was reported in a Ti-24Zr-10Nb-2Sn alloy [40].

Shape memory effect and superelasticity were also confirmed in Ti-Zr-Mo-Sn alloys. Fig. 2.27 shows the stress-strain curves of (Ti-Zr)-Mo-Sn alloys, where the atomic ratio of Ti/Zr is fixed at 1, measured by loading-unloading tensile tests at room temperature [41]. The (Ti-Zr)-1.5Mo-2Sn alloy exhibited shape memory effect, and the (Ti-Zr)-1.5Mo-3Sn and (Ti-Zr)-2Mo-2Sn alloys exhibited superelasticity. The superelastic properties were sensitively dependent on the annealing temperature. As shown in Fig. 2.28 [41], the maximum recovery strain of the (Ti-Zr)-1.5Mo-3Sn alloy increased with increasing annealing temperature. A large recovery strain of 7.0% was obtained for the specimen heat treated at 1073 K. The increase in the recovery strain was confirmed due to the large transformation strain and development of a strong $\{100\}\langle 011\rangle$ recrystallization texture.

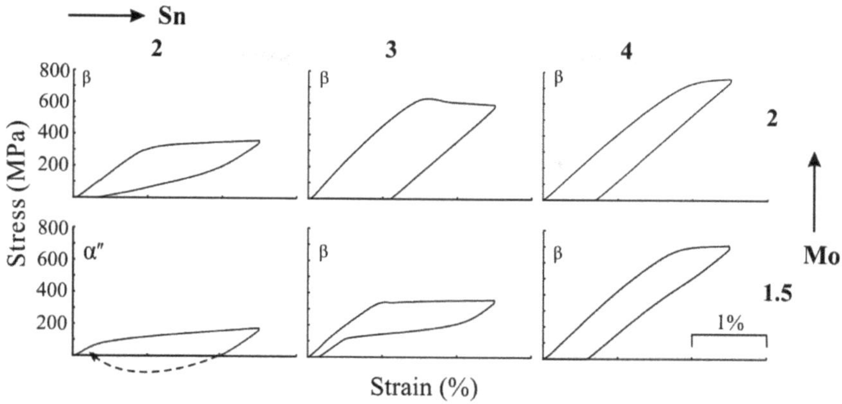

Fig. 2.27 Stress-strain curves obtained by loading-unloading tensile tests for (Ti-Zr)-Mo-Sn alloys.

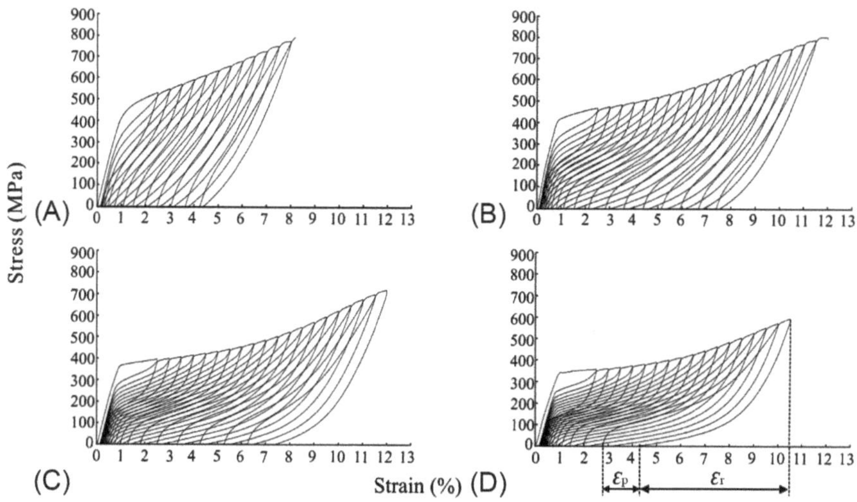

Fig. 2.28 Stress-strain curves obtained from strain increment cyclic tensile tests of a (Ti-Zr)-1.5Mo-3Sn specimens heat treated at (A) 873, (B) 923, (C) 973, and (D) 1073K for 3.6ks.

2.6 SUMMARY

Over the last decade, there have been significant efforts not only to develop new biomedical shape memory and superelastic alloys but also to understand the martensitic transformation mechanism of Ti-based alloys. Particularly, Ti-Nb-based alloys have attracted extensive research interest. The effects of alloying elements on the martensitic transformation temperatures and transformation strain of Ti-Nb-based alloys have been clarified, and many alloy systems have been developed. However, most of Ti-Nb-based alloys exhibit a relatively small superelastic recovery strain 3%–4% that is due to the intrinsic small transformation strain. Therefore, the increase in the transformation strain has been a critical issue. Zr is an effective alloying element for improving superelastic strain because Zr has weak effect on the transformation strain although it decreases the martensitic transformation temperature. The addition of Mo has similar effect on the transformation strain. However, the decrease in the amount of Nb causes the increase of the athermal ω phase, leading to the increase in the critical stress for martensitic transformation. The increase of athermal ω phase also increases the stress hysteresis and deteriorates the superelastic properties. The addition of Sn and Al is effective to suppress the formation of the athermal ω phase and improves the superelastic properties. Ti-Zr-Nb-Sn alloys exhibit a large superelastic strain of 5% owing to the combination of beneficial effects of Zr and Sn. Ti-Ta-based alloys are promising candidates for high-temperature shape memory alloys. However, the stability of Ti-Ta-based alloys is not enough. Improvement of thermal stability is a key issue in order to enable practical application as high-temperature shape memory alloys.

REFERENCES

[1] C. Baker, The shape-memory effect in a Titanium–35 wt.% niobium alloy, J. Metal Sci. 5 (1971) 92–100.

[2] T.W. Duerig, J. Albrecht, D. Richter, P. Fischer, Formation and reversion of stress induced martensite in Ti–10V–2Fe–3Al, Acta Metall. 30 (1982) 2161–2172.

[3] H. Sasano, T. Suzuki, in: G. Lutjering, U. Zwicker, W. Bunk (Eds.), Shape memory effect in Ti–Mo–Al alloys, Proceedings of the 5th International Conference on Titanium, September 10–14, 1984, Frankfurt, Germany, Deutsche Gesellschaft fur Metallkunde, 1985, pp. 1667–1674.

[4] C.Y. Lei, M.H. Wu, L.M.D. Schetky, C.J. Burstone, in: K. Inoue et al., (Eds.), Development of pseudoelastic beta titanium orthodontic wires, Proceedings of International Conference on Displacive Phase Transformations and Their Applications in Materials Engineering, May 8–9, Warrendale, PA, TMS Publications, 1998, pp. 413–418.

[5] T. Grosdidier, M.J. Philippe, Deformation induced martensite and superelasticity in a β-metastable titanium alloy, Mater. Sci. Eng. A 291 (2000) 218–223.

[6] H.Y. Kim, H. Satoru, J.I. Kim, H. Hosoda, S. Miyazaki, Mechanical properties and shape memory behavior of Ti–Nb alloys, Mater. Trans. 45 (2004) 2443–2448.

[7] H.Y. Kim, Y. Ikehara, J.I. Kim, H. Hosoda, S. Miyazaki, Martensitic transformation, shape memory effect and superelasticity of Ti–Nb binary alloys, Acta Mater. 54 (2006) 2419–2429.

[8] K. Nitta, S. Watanabe, N. Masahashi, H. Hosoda, S. Hanada, Ni-free Ti–Nb–Sn shape memory alloys, in: M. Niinomi et al., (Eds.), Structural Biomaterials for the 21st Century, TMS, Warrendale, PA, 2001, pp. 25–34.

[9] E. Takahashi, T. Sakurai, S. Watanabe, N. Masahashi, S. Hanada, Effect of heat treatment and Sn content on superelasticity in biocompatible TiNbSn alloys, Mater. Trans. 43 (2002) 2978–2983.
[10] H.Y. Kim, S. Hashimoto, J.I. Kim, T. Inamura, H. Hosoda, S. Miyazaki, Effect of Ta addition on shape memory behavior of Ti–22Nb alloy, Mater. Sci. Eng. A 417 (2006) 120–128.
[11] J.I. Kim, H.Y. Kim, T. Inamura, H. Hosoda, S. Miyazaki, Shape memory characteristics of Ti–22Nb–(2-8)Zr(at.%) biomedical alloys, Mater. Sci. Eng. A 403 (2005) 334–339.
[12] Y. Al-Zain, H.Y. Kim, H. Hosoda, T.H. Nam, S. Miyazaki, Shape memory properties of Ti–Nb–Mo biomedical alloys, Acta Mater. 58 (2010) 4212–4223.
[13] B.L. Wang, Y.F. Zheng, L.C. Zhao, Effects of Sn content on the microstructure, phase constitution and shape memory effect of Ti–Nb–Sn alloys, Mater. Sci. Eng. A 486 (2008) 146–151.
[14] Y. Fukui, T. Inamura, H. Hosoda, K. Wakashima, S. Miyazaki, Mechanical properties of a Ti–Nb–Al shape memory alloy, Mater. Trans. 45 (2004) 1077–1082.
[15] K. Masumoto, Y. Horiuchi, T. Inamura, H. Hosoda, K. Wakashima, H.Y. Kim, S. Miyazaki, Effect of Si addition on superelastic properties of Ti–Nb–Al biomedical shape memory alloys, Mater. Sci. Eng. A 438–440 (2006) 835–838.
[16] T. Inamura, Y. Fukui, H. Hosoda, K. Wakashima, S. Miyazaki, Mechanical properties of Ti–Nb biomedical shape memory alloys containing Ge or Ga, Mater. Sci. Eng. C 25 (2005) 426–432.
[17] H.Y. Kim, T. Sasaki, K. Okutsu, J.I. Kim, T. Inamura, H. Hosoda, S. Miyazaki, Texture and shape memory behavior of Ti–22Nb–6Ta alloy, Acta Mater. 54 (2006) 423–433.
[18] E. Bertrand, T. Gloriant, D.M. Gordin, E. Vasilescu, P. Drob, C. Vasilescu, S.I. Drob, Synthesis and characterization of a new superelastic Ti–25Ta–25Nb biomedical alloy, J. Mech. Behav. Biomed. Mater. 3 (2010) 559–564.
[19] J.I. Kim, H.Y. Kim, T. Inamura, H. Hosoda, S. Miyazaki, Effect of annealing temperature on microstructure and shape memory characteristics of Ti–22Nb–6Zr(at%) biomedical alloy, Mater. Trans. (3) (2006) 505–512.
[20] V. Brailovski, S. Prokoshkin, M. Gauthier, K. Inaekyan, S. Dubinskiy, M. Petrzhik, M. Filonov, Bulk and porous metastable beta Ti–Nb–Zr(ta) alloys for biomedical applications, Mater. Sci. Eng. C 31 (2011) 643–657.
[21] F. Sun, Y.L. Hao, S. Nowak, T. Gloriant, P. Laheurte, F. Prima, A thermal-mechanical treatment to improve the superelastic performances of biomedical Ti–26Nb and Ti–20Nb–6Zr(at.%) alloys, J. Mech. Behav. Biomed. Mater. 4 (2011) 1864–1872.
[22] Q. Li, M. Niinomi, M. Nakai, Z.D. Cui, S.L. Zhu, X.J. Yang, Effect of Zr on superelasticity and mechanical properties of Ti–24 at% Nb–(0, 2, 4) at% Zr alloy subjected to aging treatment, Mater. Sci. Eng. A 536 (2012) 197–206.
[23] J.Y. Zhang, F. Sun, Y.L. Hao, N. Gozdecki, E. Lebrun, P. Vermaut, R. Portier, T. Gloriant, P. Laheurte, F. Prima, Influence of equiatomic Zr/Nb substitution on superelastic behavior of Ti–Nb–Zr alloy, Mater. Sci. Eng. A 563 (2013) 78–85.
[24] J.I. Kim, H.Y. Kim, H. Hosoda, S. Miyazaki, Shape memory behavior of Ti–22Nb–(0.5-2.0)O(at%) biomedical alloys, Mater. Trans. 46 (2005) 852–857.
[25] M. Tahara, H.Y. Kim, H. Hosoda, S. Miyazaki, Shape memory effect and cyclic deformation behavior of Ti–Nb–N alloys, Funct. Mater. Lett. 2 (2009) 79–82.
[26] A. Ramarolahy, P. Castany, F. Prima, P. Laheurte, I. Peron, T. Gloriant, Microstructure and mechanical behavior of superelastic Ti–24Nb–0.5O and Ti–24Nb–0.5N biomedical alloys, J. Mech. Behav. Biomed. Mater. 9 (2012) 83–90.
[27] H.Y. Kim, N. Oshika, J.I. Kim, T. Inamura, H. Hosoda, S. Miyazaki, Martensitic transformation and superelasticity of Ti–Nb–Pt alloys, Mater. Trans. 48 (2007) 400–406.

[28] D.H. Ping, Y. Mitarai, F.X. Yin, Microstructure and shape memory behavior of a Ti–30Nb–3Pd alloy, Scr. Mater. 52 (2005) 1287–1291.
[29] N. Sakaguch, M. Niinomi, T. Akahori, Tensile deformation behavior of Ti–Nb–ta–Zr biomedical alloys, Mater. Trans. 45 (2004) 1113–1119.
[30] L.Q. Wang, W.J. Lu, J.N. Qin, F. Zhang, D. Zhang, Texture and superelastic behavior of cold-rolled TiNbTaZr alloy, Mater. Sci. Eng. A 491 (2008) 372–377.
[31] Y.L. Hao, S.J. Li, S.Y. Sun, R. Yang, Effect of Zr and Sn on Young's modulus and superelasticity of Ti–Nb-based alloys, Mater. Sci. Eng. A 441 (2006) 112–118.
[32] E.G. Obbard, Y.L. Hao, R.J. Talling, S.J. Li, Y.W. Zhang, D. Dye, R. Yang, The effect of oxygen on α″ martensite and superelasticity in Ti–24Nb–4Zr–8Sn, Acta Mater. 59 (2011) 112–125.
[33] D.C. Zhang, S. Yang, M. Wei, Y.F. Mao, C.G. Tan, J.G. Lin, Effect of Sn addition on the microstructure and superelasticity in Ti–Nb–Mo–Sn alloys, J. Mech. Behav. Biomed. Mater. 13 (2012) 156–165.
[34] M.F. Ijaz, H.Y. Kim, H. Hosoda, S. Miyazaki, Effect of Sn addition on stress hysteresis and superelastic properties of a Ti–15Nb–3Mo alloy, Scr. Mater. 72–73 (2014) 29–32.
[35] H. Tada, T. Yamamoto, X.M. Wang, H. Kato, Effect of Al addition on superelastic properties of aged Ti–Nb–Zr–Al quaternary alloys, Mater. Trans. 53 (2012) 1981–1985.
[36] M. Tahara, H.Y. Kim, H. Hosoda, T.Y. Nam, S. Miyazaki, Effect of nitrogen addition and annealing temperature on superelastic properties of Ti–Nb–Zr–Ta alloys, Mater. Sci. Eng. A 527 (2010) 6844–6852.
[37] D. Kent, G. Wang, Z. Yu, M.S. Dargusch, Pseudoelastic behaviour of a β Ti–25Nb–3Zr–3Mo–2Sn alloy, Mater. Sci. Eng. A 527 (2010) 2246–2252.
[38] Y. Cui, Y. Li, K. Luo, H.B. Xu, Microstructure and shape memory effect of Ti–20Zr–10Nb alloy, Mater. Sci. Eng. A 527 (2010) 652–656.
[39] J. Fu, A. Yamamoto, H.Y. Kim, H. Hosoda, S. Miyazaki, Novel Ti-base superelastic alloys with large recovery strain and excellent biocompatibility, Acta Biomater. 17 (2015) 56–67.
[40] L.L. Pavón, H.Y. Kim, H. Hosoda, S. Miyazaki, Effect of Nb content and heat treatment temperature on superelastic properties of Ti–24Zr–(8–12)Nb–2Sn alloys, Scr. Mater. 95 (2015) 46–49.
[41] M.F. Ijaz, H.Y. Kim, H. Hosoda, S. Miyazaki, Superelastic properties of biomedical (Ti–Zr)–Mo–Sn alloys, Mater. Sci. Eng. C 48 (2014) 11–20.
[42] P.J.S. Buenconsejo, H.Y. Kim, S. Miyazaki, Shape memory behavior of Ti–Ta and its potential as a high-temperature shape memory alloy, Acta Mater. 57 (2009) 1068–1077.
[43] P.J.S. Buenconsejo, H.Y. Kim, S. Miyazaki, Effect of ternary alloying elements on the shape memory behavior of Ti–Ta alloys, Acta Mater. 57 (2009) 2509–2515.
[44] H.Y. Kim, T. Fukushima, P.J.S. Buenconsejo, T.H. Nam, S. Miyazaki, Martensitic transformation and shape memory properties of Ti–Ta–Sn high temperature shape memory alloys, Mater. Sci. Eng. 528 (2011) 7238–7246.
[45] P.J.S. Buenconsejo, H.Y. Kim, S. Miyazaki, Novel beta-TiTaAl alloys with excellent cold workability and a stable high-temperature shape memory effect, Scr. Mater. 64 (2011) 1114–1117.
[46] H.Y. Kim, T. Kanaya, T. Fukushima, P.J.S. Buenconsejo, S. Miyazaki, Stability of Ti-Ta base high temperature shape memory alloys, Mater. Sci. For. 706–709 (2012) 1921–1924.
[47] H.Y. Kim, S. Miyazaki, Martensitic transformation and superelastic properties of Ti–Nb base alloys, Mater. Trans. 56 (2015) 625–634.
[48] H.Y. Kim, J. Fu, H. Tobe, J.I. Kim, S. Miyazaki, Crystal structure, transformation strain, and superelastic property of Ti–Nb–Zr and Ti–Nb–Ta alloys, Shape Mem. Superelasticity 1 (2015) 107–116.

[49] S. Miyazaki, H.Y. Kim, H. Hosoda, Development and characterization of Ni-free Ti-base shape memory and superelastic alloys, Mater. Sci. Eng. A 438–440 (2006) 18–24.
[50] Y. Al-Zain, H.Y. Kim, T. Koyano, H. Hosoda, T.H. Nam, S. Miyazaki, Anomalous temperature dependence of the superelastic behavior of Ti–Nb–Mo alloys, Acta Mater. 59 (2011) 1464–1473.
[51] W.H. Ho, C.P. Ju, J.H.C. Lin, Structure and properties of cast binary Ti–Mo alloys, Biomaterials 20 (1999) 2115–2122.
[52] T. Maeshima, S. Ushimaru, K. Yamauchi, M. Nishida, Effects of Sn content and aging conditions on superelasticity in biomedical Ti–Mo–Sn alloys, Mater. Trans. 47 (2006) 513–517.
[53] H.Y. Kim, Y. Ohmatsu, J.I. Kim, H. Hosoda, S. Miyazaki, Mechanical properties and shape memory behavior of Ti–Mo–Ga alloys, Mater. Trans. 45 (2004) 1090–1095.
[54] T. Zhou, M. Aindow, S.P. Alpay, M.J. Blackburn, M.H. Wu, Pseudo-elastic deformation behavior in a Ti/Mo-based alloy, Scr. Mater. 50 (2004) 343–348.
[55] H.Y. Kim, Y. Ohmatsu, J.I. Kim, T. Inamura, H. Hosoda, S. Miyazaki, Effect of Nb addition on shape memory behavior of Ti–Mo–Ga alloys, Mater. Trans. 47 (2006) 518–522.
[56] T. Maeshima, M. Nishida, Shape memory properties of biomedical Ti–Mo–Ag and Ti–Mo–Sn alloys, Mater. Trans. 45 (2004) 1096–1100.
[57] T. Maeshima, M. Nishida, Shape memory and mechanical properties of biomedical Ti–Sc–Mo alloys, Mater. Trans. 45 (2004) 1101–1105.
[58] Y. Sutou, K. Yamauchi, T. Takagi, T. Maeshima, M. Nishida, Mechanical properties of Ti–6 at.% Mo–4 at.% Sn alloy wires and their application to medical guidewire, Mater. Sci. Eng. A 438–440 (2006) 1097–1100.

CHAPTER 3

Effect of Interstitial Alloying Elements on Shape Memory and Superelastic Properties

Contents

3.1 INTRODUCTION

Mechanical properties of pure Ti and Ti alloys are strongly influenced by the interstitial alloying elements [1–4]. Oxygen has attracted particular attention not only because of its strong hardening effect but also because of its technological importance since it is an unavoidable impurity in Ti and Ti alloys. Concentration of interstitial alloying elements has also been considered as an important parameter to determine the deformation mode of Ti and Ti alloys [5–8].

The interstitial alloying elements also significantly affect the phase stability and shape memory properties of Ti alloys [9–13]. Furthermore, it has been reported that the interstitial alloying elements cause many unique and anomalous properties such as low Young's modulus, nonlinear elastic behavior, high damping capacity, and abnormal temperature dependence of shape memory behavior [14–27]. The interstitial alloying elements also induce a unique microstructure, i.e., a nanosized lattice modulation (nanodomain) structure [22–27].

Ni-free Ti-based Shape Memory Alloys
https://doi.org/10.1016/B978-0-12-809401-3.00003-3

In this chapter, the effects of interstitial alloying elements such as oxygen, nitrogen, boron, and carbon on microstructure, mechanical properties, and transformation behavior of metastable Ti-Nb alloys are reviewed.

3.2 OXYGEN

Oxygen is the most common and unavoidable impurity in titanium and titanium alloys and affects their mechanical properties such as strength, ductility, toughness, and fatigue properties. It has been found that the amount of oxygen also influences the martensitic transformation behavior and shape memory properties [9, 12, 13, 28]. Fig. 3.1 shows the effect of oxygen addition on the stress-strain curves of Ti-Nb binary alloys [28]. The specimens were elongated to the strain of 2.5% and unloaded. After unloading, the specimens were heated up to 500K to examine shape memory effect: dashed lines with an arrow indicate shape recovery upon heating. As explained in Chapter 2, the Ti-Nb binary alloys exhibit shape memory effect and superelasticity in Ti-(20–25)Nb and Ti-(26, 27)Nb alloys, respectively. For the Ti-Nb-1O alloys, superelastic recovery is observed when the Nb content is 22–24at.%. With increasing Nb content, the stress hysteresis decreases, and then, nonlinear elastic behavior is observed in Ti-(25, 26)Nb-1O alloys. On the other hand, Ti-(27, 28)Nb-1O alloys exhibit permanent deformation.

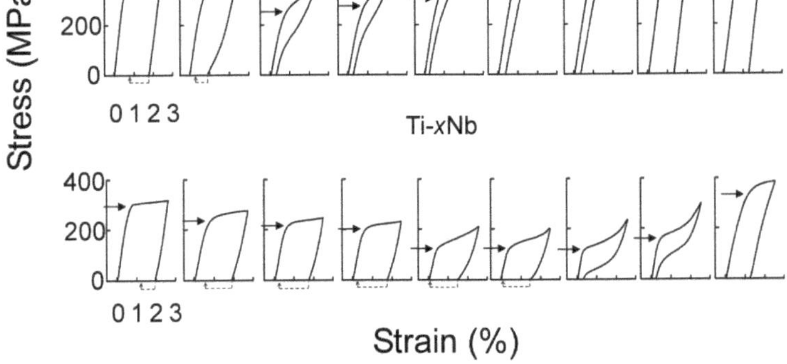

Fig. 3.1 Stress-strain curves obtained by loading-unloading tensile tests for Ti-(20–28)Nb and Ti-(20–28)Nb-1O alloys.

An arrow on each stress-strain curve in Fig. 3.1 indicates the point at which apparent yielding occurs. The apparent yield stress is plotted in Fig. 3.2 as a function of the Nb content. For the binary Ti-Nb alloys, the apparent yield stress decreases with increasing Nb content, reaching a minimum value at 26 at.% Nb and then increases again. The deformation mechanism of apparent yielding depends on the transformation temperature that is dependent on the Nb content. For the Ti-(20–24)Nb alloys, the apparent yielding is due to the reorientation of martensite variants by twinning because the martensitic transformation finish temperature of the alloys is above room temperature and alloys are mainly composed of self-accommodating martensite variants. Therefore, for the Ti-(20–24)Nb alloys, the apparent yield stress corresponds to the critical stress for reorientation of martensite variants. For the Ti-(26, 27)Nb alloys, the yield occurs through the stress-induced martensitic transformation because the martensitic transformation temperature is slightly lower than room temperature. The critical stress for inducing martensitic transformation of the Ti-27Nb alloy is higher than that of the Ti-26Nb alloy because of its lower martensitic transformation temperature. For the Ti-28Nb alloy, the martensitic transformation temperature is too low to induce the stress-induced martensitic transformation at room temperature; thus, apparent yielding is due to the

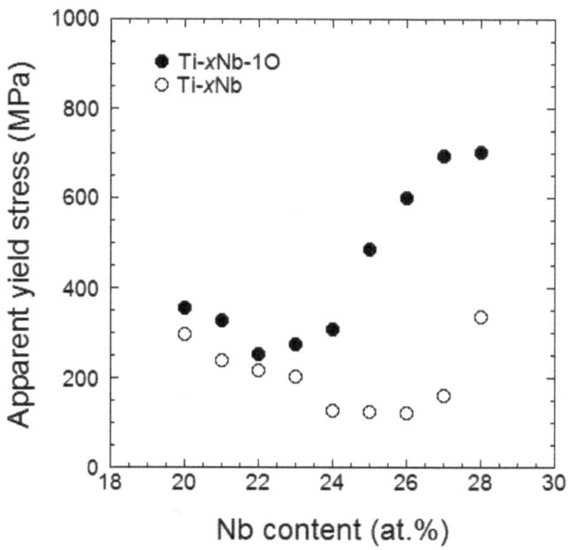

Fig. 3.2 Nb content dependence of the apparent yield stress for Ti-(20–28)Nb and Ti-(20–28)Nb-1O alloys.

plastic deformation by dislocation movement or twinning because the plastic deformation occurs before the onset of stress-induced martensitic transformation. A similar composition dependence of the apparent yield stress is also observed in oxygen-added alloys: the apparent yield stress decreases with increasing Nb content until a minimum value is reached at 22 at.% and then increases with further increasing Nb content. Superelastic recovery is observed in the Ti-(22–24)Nb-1O alloys. It is noted that the Nb content exhibiting the minimum yield stress is 4 at.% lower than that in the Ti-Nb binary alloys. It is also seen that the critical stress for inducing martensitic transformation increases by the addition of oxygen. From the comparison of the Nb contents of both alloys exhibiting superelasticity with similar critical stresses for inducing martensitic transformation, it is suggested that the 1 at.% oxygen addition decreases the martensitic transformation temperature by about 160–200 K because the martensitic transformation is reported to decrease with a slope of −40 K/at.% Nb for Ti-(20–28)Nb alloys. It should be mentioned that the addition of oxygen increases the critical stress for plastic deformation: the yield stress of the Ti-28Nb alloy increases from 350 to 700 MPa by the addition of 1 at.% oxygen. Fig. 3.3

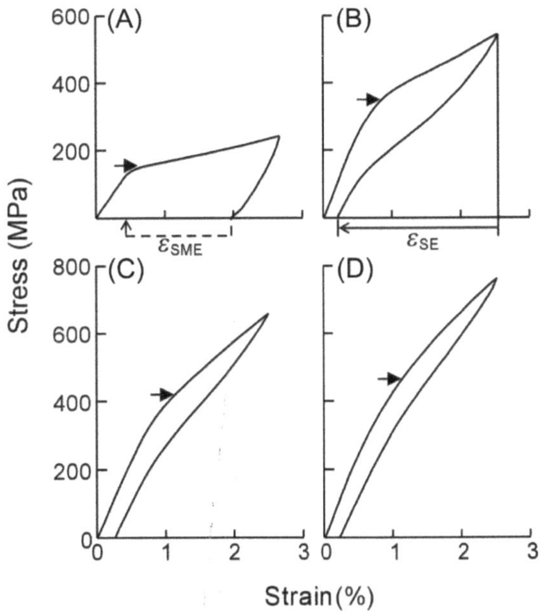

Fig. 3.3 Effect of O content on the stress-strain curves Ti-22Nb-(0.5–2)O alloys upon loading and unloading. (A) Ti-22Nb-0.5O, (B) Ti-22Nb-1O, (C) Ti-22Nb-1.5O, and (D) Ti-22Nb-2O.

shows the effect of oxygen concentration on shape memory properties [12]. As the oxygen concentration increases, the stress for inducing martensitic transformation increases, and the deformation behavior changes from shape memory effect to superelasticity and then to nonlinear elastic behavior, confirming that oxygen decreases the transformation temperatures.

The effect of the oxygen addition on deformation behavior is summarized as follows (see Fig. 3.4): (1) The critical stress for plastic deformation (σ_{pd}) increases, and (2) the martensitic transformation temperature decreases. The decrease in the martensitic transformation temperature causes superelasticity to occur at a lower Nb content. On the other hand, the increase in the critical stress for plastic deformation allows that superelasticity occurs at a higher Nb content. As a result, superelasticity occurs over a wider composition range in oxygen-added alloys. Furthermore, the addition of oxygen leads to a larger transformation strain at compositions showing superelastic recovery because the transformation strain increases as the Nb content decreases.

3.3 NITROGEN

Fig. 3.5 shows the effect of nitrogen addition on the deformation behavior of Ti-Nb alloys. The impact of the addition of nitrogen on the deformation behavior of Ti-Nb alloys is similar to that of oxygen. Superelastic recovery is observed in the alloys with lower Nb content compared with the Ti-Nb binary alloys: superelasticity is observed at 23–25 at.% Nb for the 0.5N-added alloys and 21–25 at.% Nb for the 1N-added alloys, respectively. It is also noted that the composition range exhibiting superelasticity becomes

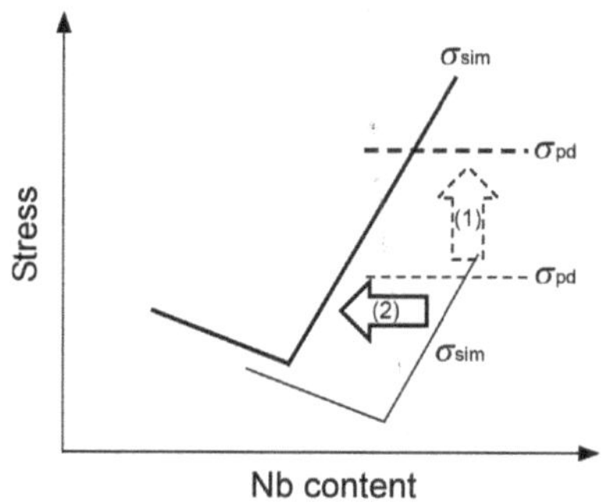

Fig. 3.4 A schematic illustration explaining the effect of the oxygen addition on the critical stress for inducing martensitic transformation and the stress for plastic deformation.

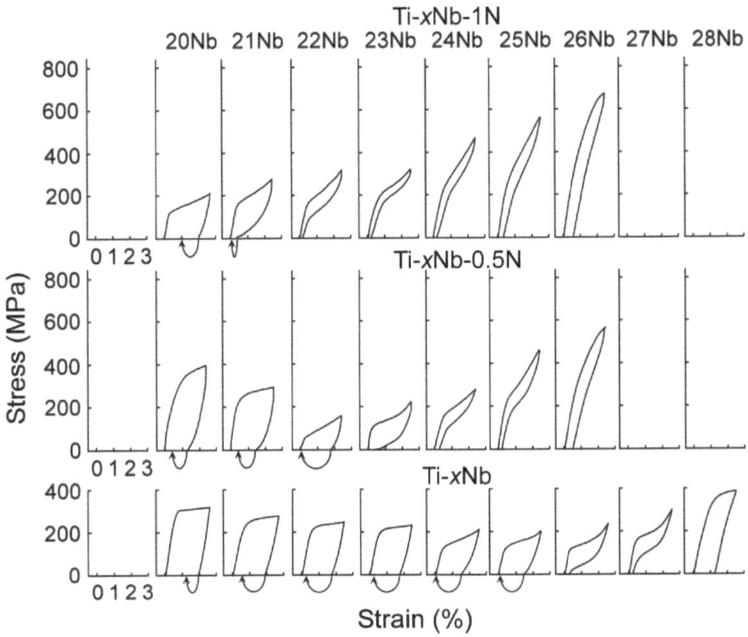

Fig. 3.5 Stress-strain curves obtained by loading-unloading tensile tests for Ti-(20–28) Nb and Ti-(20–28)Nb-1O alloys.

wider by the addition of N. The apparent yield stress decreases with increasing Nb content until reaching a minimum value and then increases with further increasing Nb content. The minimum yield stress is observed at 25 at.% Nb for the binary alloys, 22 at.% Nb for the 0.5N-added alloys, and 20 at.% Nb for the 1N-added alloys, respectively, implying that the impact of 1 at.% N for reducing the martensitic transformation temperature is equivalent to about 5 at.% Nb. This suggests that the 1 at.% N addition decreases the martensitic transformation temperature by about 160–200 K [29].

The addition of nitrogen drastically improves the superelastic properties in terms of recovery strain, critical stress for plastic deformation, and cyclic stability. Fig. 3.6 compares superelastic properties of Ti-27Nb and Ti-23Nb-1N alloys [28]; both alloys exhibit the most excellent superelastic properties among Ti-Nb binary alloys and Ti-Nb-1N alloys, respectively. The superelastic recovery strain increases from 2.0% to 3.6% by the addition of 1 at.% N. The large superelastic recovery strain in the Ti-23Nb-1N alloy is attributed to two major reasons: One is the increase in the critical stress for plastic deformation, and another is the increase in the intrinsic transformation strain. It is clearly seen that the addition of 1 at.% N increases the secondary yielding stress from 350 to 700 MPa corresponding to the critical

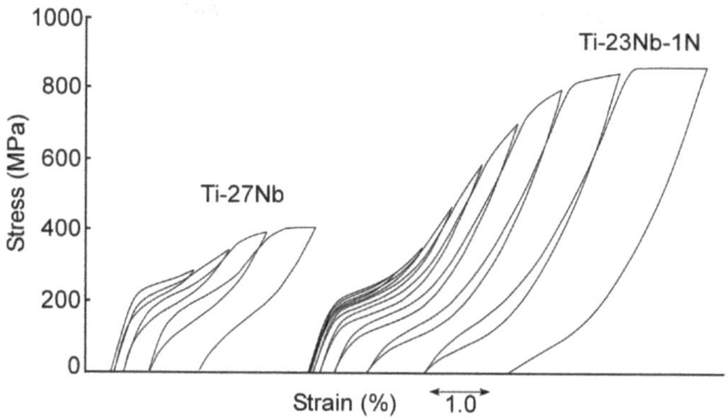

Fig. 3.6 Stress-strain curves obtained by strain increment cyclic loading and unloading tensile tests for Ti-27Nb and Ti-23Nb-1N alloys.

stress for plastic deformation of stress-induced martensite phase, although the first yielding stress corresponding to the critical stress for inducing martensitic transformation remains similar. A higher critical stress for plastic deformation is preferable for the superelastic recovery strain because it prevents permanent deformation during stress-induced martensitic transformation. The addition of N decreases martensitic transformation temperature, which allows superelasticity at lower Nb content. Therefore, the addition of N increases the intrinsic transformation strain because the transformation strain increases with decreasing Nb content as explained in Chapter 2.

The positive effect of the N addition on superelastic properties has been confirmed in other Ti-Nb-based alloys [30]. Fig. 3.7 shows the effect of the

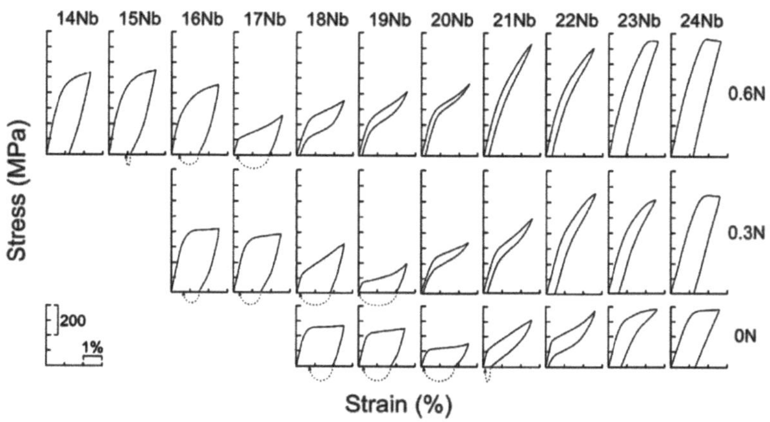

Fig. 3.7 Stress-strain curves obtained by loading-unloading tensile tests for Ti-(18–24)Nb-4Zr-2Ta, Ti-(16–24)Nb-4Zr-2Ta-0.3N, and Ti-(14–24)Nb-4Zr-2Ta-0.6N alloys.

N addition on the shape memory properties in Ti-Nb-4Zr-2Ta alloys [30]. The apparent yield stress takes a minimum at the Nb content of 20 at.%, while it takes a minimum at 19 and 17 at.% in 0.3N and 0.6N-added alloys. For the Ti-Nb-4Zr-2Ta alloys without N addition, superelasticity is observed when the Nb content is 21–23 at.%. Similarly to the Ti-Nb binary alloys, the addition of N decreases the Nb content of the alloys showing superelasticity, i.e., 20–23 at.% in the 0.3N-added alloys and 18–22 at.% in the 0.6N-added alloys. These results confirm that the addition of N decreases the martensitic transformation temperatures.

Fig. 3.8 shows the effect of deformation temperature on the stress-strain curve for Ti-20Nb-4Zr-2Ta-0.3N and Ti-20Nb-4Zr-2Ta-0.6N alloys [30]. It is seen that the both alloys exhibit superelasticity in wide temperature ranges. The apparent yield stress corresponding to the critical stress for inducing martensitic transformation is plotted as a function of deformation temperature for Ti-20Nb-4Zr-2Ta-0.3N and Ti-20Nb-4Zr-2Ta-0.6N alloys in Fig. 3.9 [30]. The apparent yield stress increases with increasing deformation temperature in accordance with the Clausius-Clapeyron relationship of the stress-induced martensitic transformation: $d\sigma/dT = -\Delta H/T\cdot\varepsilon = -\Delta S/\varepsilon$, where σ is a critical stress for inducing martensitic transformation, ε is a transformation strain, and ΔH and ΔS are the enthalpy and entropy of the martensitic transformation per unit volume, respectively. It is seen that the linear regression lines of both alloys are almost parallel and the line of 0.6N alloy is overlapped with the line of 0.3N alloy by shifting to 60 K higher temperature, implying that the difference of the

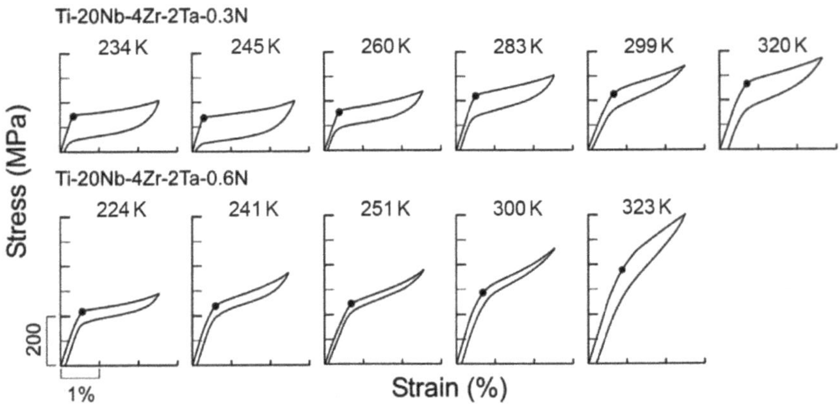

Fig. 3.8 Effect of deformation temperature on the stress-strain curve for a Ti-20Nb-4Zr-2Ta-0.3N alloy upon loading and unloading.

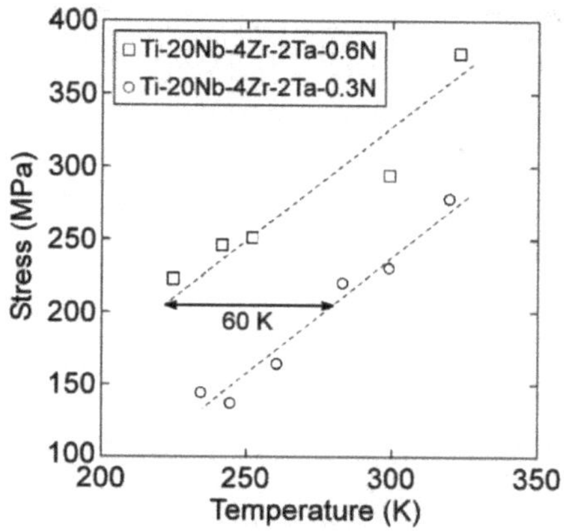

Fig. 3.9 Deformation temperature dependence of the critical stress for inducing martensitic transformation for Ti-(16–24)Nb-4Zr-2Ta-0.3N and Ti-(14–24)Nb-4Zr-2Ta-0.6N alloys.

martensitic transformation temperatures of the two alloys is about 60 K. As a result, it is concluded that the addition of 1 at.% N decreases the martensitic transformation temperature by about 200 K. It is also noted in Fig. 3.8 that the stress hysteresis of the 0.6N-added alloy is smaller than that of the 0.3N-added alloy.

The addition of N is also effective on improving superelastic properties of Ti-Zr-based ternary alloys. Fig. 3.10 shows the results of strain increment

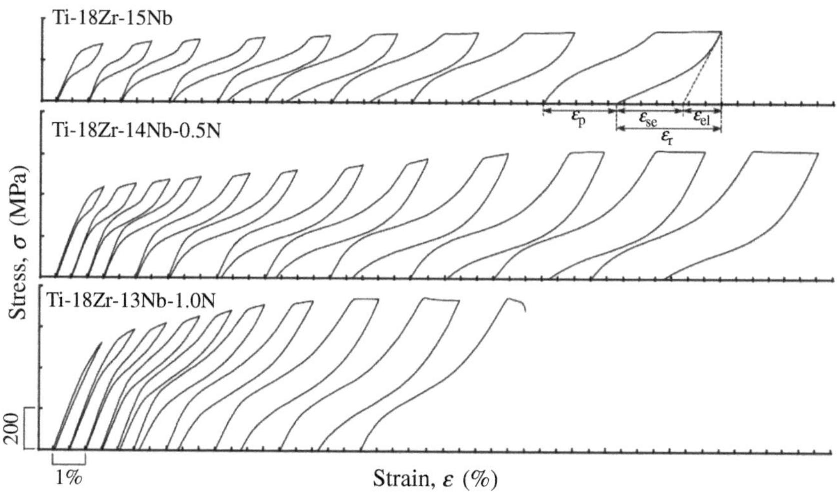

Fig. 3.10 Stress-strain curves obtained by strain increment cyclic loading and unloading tensile tests for Ti-18Zr-15Nb, Ti-18Zr-14Nb-0.5N, and Ti-18Zr-13Nb-1N alloys.

cyclic tensile tests of Ti-18Zr-15Nb, Ti-18Zr-14Nb-0.5N, and Ti-18Zr-13Nb-1N alloys [31]. The maximum recovery strain increases from 3.5% to 5.0% and 4.5% by the addition of 0.5N and 1N, respectively, due to the increase of the critical stress for plastic deformation. However, the increase in the N content leads to the loss of ductility. It is also noted that the strengthening effect in Ti-Zr-Nb alloys is relatively weaker than that in Ti-Nb binary alloys. This weaker effect of N on the critical stress for plastic deformation in the Ti-Zr-Nb alloys is suggested to be due to their larger lattice parameter of the β phase when compared with Ti-Nb binary alloys.

3.4 BORON AND CARBON

Fig. 3.11 shows the effect of boron addition on the deformation behavior of Ti-Nb binary alloys. It is seen that the addition of B affects the shape

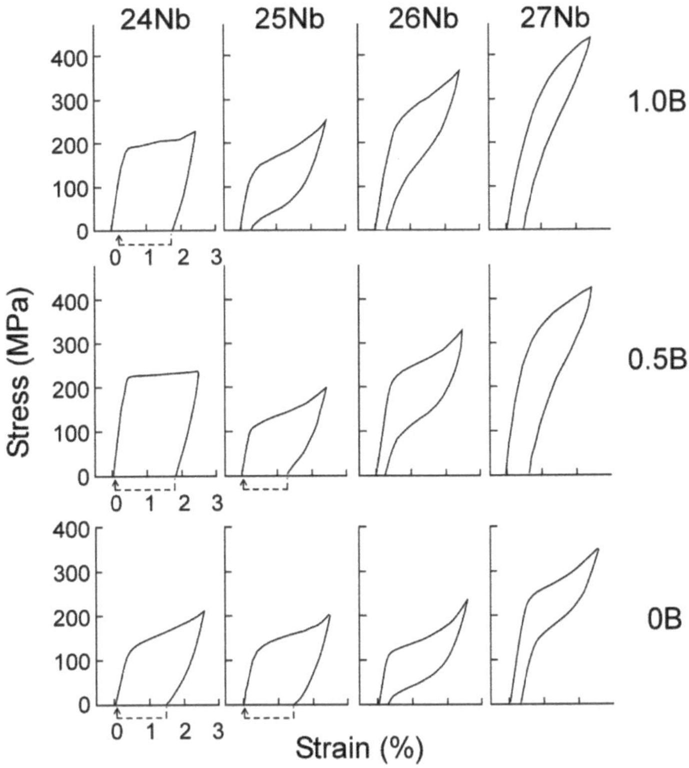

Fig. 3.11 Stress-strain curves obtained by loading-unloading tensile tests for Ti-(24–27)Nb, Ti-(24–27)Nb-0.5B, and Ti-(14–27)Nb-1B alloys.

memory properties of Ti-Nb alloys although the impact of the addition of boron on the deformation behavior is not as drastic as those of the oxygen and nitrogen. The addition of 1 at.% B causes superelasticity in the Ti-25Nb alloy. And the stress-induced martensitic transformation of the Ti-26Nb alloy increases slightly with increasing B content of the alloy. These results indicate that B decreases the martensitic transformation temperature although the amount of decrease is smaller when compared with O and N. This weak effect of the B addition on the shape memory properties of Ti-Nb alloys is due to the low solubility of B in the β phase. Fig. 3.12 shows the bright-field TEM images of Ti-26Nb-(0–0.7)B alloys heat treated at 1173 K for 1.8 ks [32]. The Ti-26Nb and Ti-26Nb-0.1B alloys are composed only the β phase, while the precipitates of TiB are observed in the

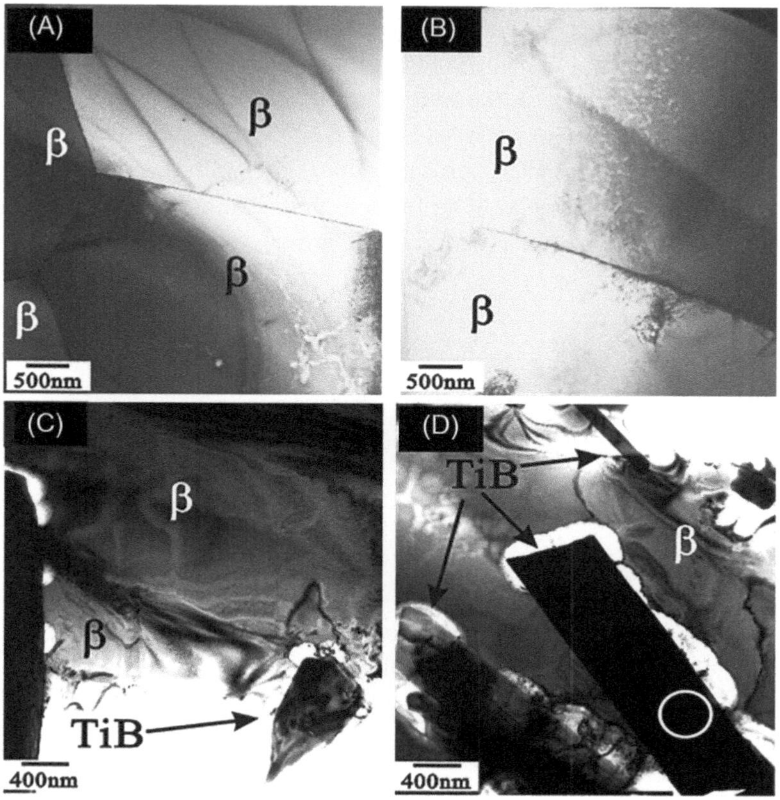

Fig. 3.12 Bright-field TEM images of (A) Ti-26Nb, (B) Ti-26Nb-0.1B, (C) Ti-26Nb-0.2B, and (D) Ti-26Nb-0.7B alloys heat treated at 1173 K for 1.8 ks.

Ti-26Nb-0.2B and Ti-26Nb-0.7B alloys, implying that the solubility limit of B in the β phase of the Ti-26Nb alloy is <0.2 at.%. It should be noted that no second phases are formed in the Ti-Nb-O and Ti-Nb-N alloys when heat treated above the β transus temperature. The effect of the B addition on the martensitic transformation temperature depends on its concentration as shown in Fig. 3.13 [32]. The martensitic transformation start temperature decreases with a slope of 350 K/at.% B when B concentration is less than the solubility limit. It is noticed that this value is about two times larger than the impact of the N and O addition on the martensitic transformation temperature. The addition of B beyond the solubility limit reduces the martensitic transformation start temperature with a much smaller slope of 15 K/at.% B. This smaller effect is considered to be the precipitation effect to suppress the martensitic transformation. The addition of B raises the critical stress for plastic deformation and improves the superelastic properties; however, the strengthening effect is not appreciable due to the low solubility limit. A similar effect of the addition of B on improving the superelastic properties has been reported in a Ti-Nb-Al alloy [33, 34]. Carbon also has very limited solubility in the β phase of titanium alloys. It was reported that the TiC was formed by the addition of 0.2 wt%C in a Ti-27Nb alloy [35]. The addition of C suppresses the recrystallization and reduces the grain size due to the pinning effect of TiC particles. The refined grain structure of C-added alloy raises the critical stress for plastic deformation.

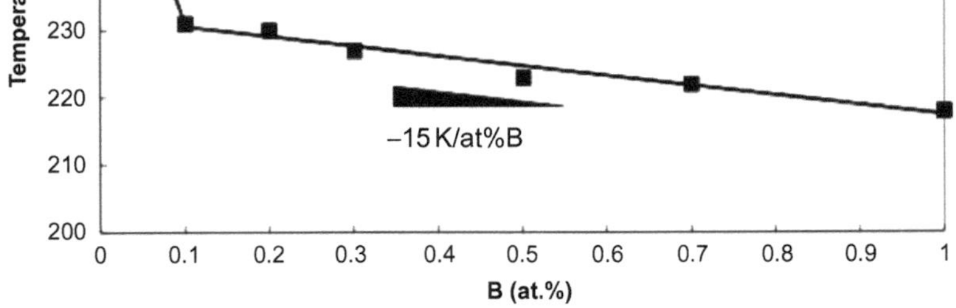

Fig. 3.13 Effect of the B content on the martensitic transformation temperature of Ti-26Nb-B alloys.

3.5 UNIQUE PROPERTIES INDUCED BY INTERSTITIAL ALLOYING ELEMENTS

3.5.1 Nano Domain Structure

It has been suggested that the suppression of the martensitic transformation by the addition of interstitial alloying elements is due to the formation of nanosized lattice modulation structure [22–31]. Fig. 3.14 shows selected area diffraction patterns of a (Ti-23Nb)-1O alloy taken from [001], [113], and [111] zone axes of the β phase [26]. A Ti-23Nb alloy exhibits a fully martensitic transformed structure at room temperature. On the other hand, no distinct martensite phase is detected in the (Ti-23Nb)-1O alloy. In three diffraction patterns of the (Ti-23Nb)-1O alloy, diffuse streaks along the $\langle 110\rangle^*$ with intensity maxima at $\left\{h+\frac{1}{2}\,k+\frac{1}{2}\,0\right\}$ positions are observed between primary diffraction spots of the β phase. These diffuse streaks are originated from $\{110\} < 1\overline{1}0 >$ type transverse atomic displacement. The diffuse streaks associated with transverse-type lattice displacement have been reported in many shape memory alloys such as Ti-Ni [36, 37], Ti-Ni-Fe [38, 39], Cu-Al-Ni [40], and Ni-Al [41].

As shown in Fig. 3.15 [23], the $\{110\}\langle 1\overline{1}0\rangle$-type lattice modulation is associated with the shuffling mode of the martensitic transformation from β to α″: the $\{110\}\langle 1\overline{1}0\rangle$-type transverse displacement is crystallographically equivalent to the shuffling of basal plane of α″ martensite phase. The $\{110\}\langle 1\overline{1}0\rangle$-type lattice modulation is divided into six variants based on their shuffling planes as given in Table 3.1 [23]. Each diffuse streak is associated with a particular transverse displacement mode, i.e., lattice modulation mode. For example, the diffuse streaks of $\left[\overline{1}10\right]^*$ and $[110]^*$ observed in Fig. 3.14A, which are associated with the lattice modulation modes of $(\overline{1}10)\langle 110\rangle$ and $(110)\langle\overline{1}10\rangle$, are corresponding to variants V5 and V6, respectively.

It has been confirmed that the six variants of lattice modulation modes are present equivalently if no external stress is applied. In real space, the lattice modulation is observed as a nanodomain structure. Fig. 3.16 shows dark-field micrographs taken from the intensity maxima at the 1/2 positions of streaks along $[110]^*$ and $\left[\overline{1}10\right]^*$ directions in a (Ti-23Nb)-1O alloy [26]. Dark-field micrographs reveal that nanometer-sized domains <3 nm are homogeneously distributed in the β-phase matrix. It is noted that there are no obvious differences in morphology of nanodomains in Fig. 3.16A and B, which are corresponding to the lattice modulation modes V6 and V5, respectively. It has been confirmed that all six variants of the lattice

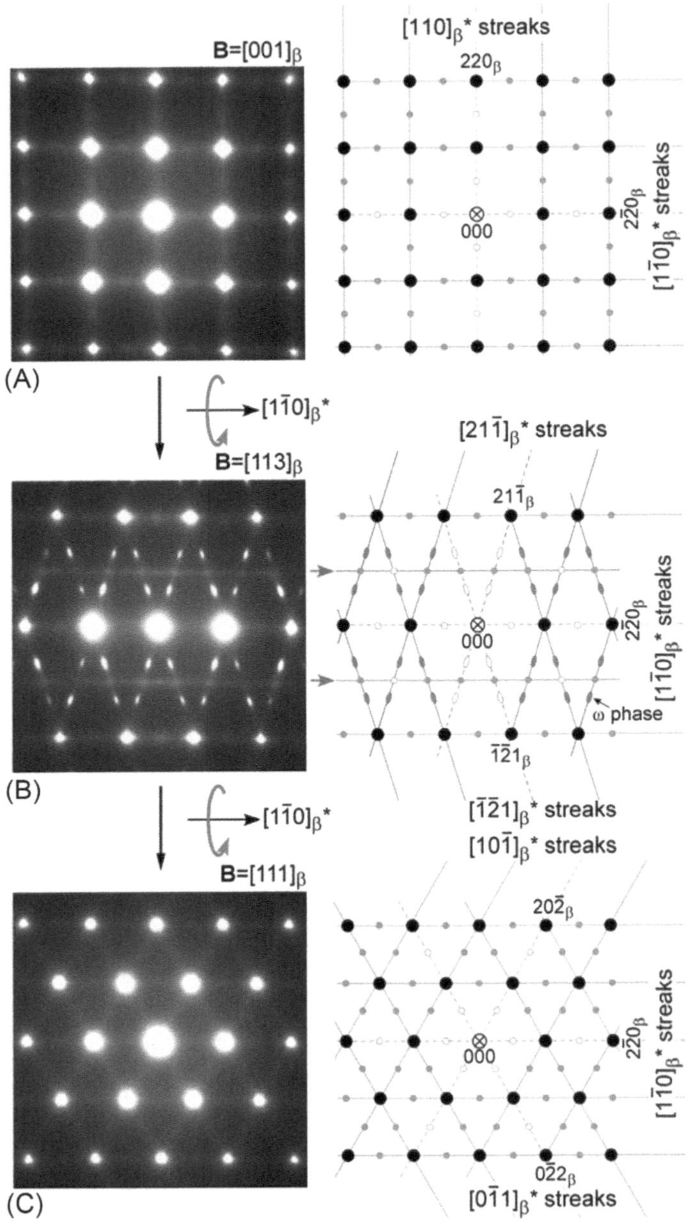

Fig. 3.14 Selected area diffraction patterns of a (Ti-23Nb)-1O alloy taken from (A) [001], (B) [113], and (C) [111] zone axis of the β phase.

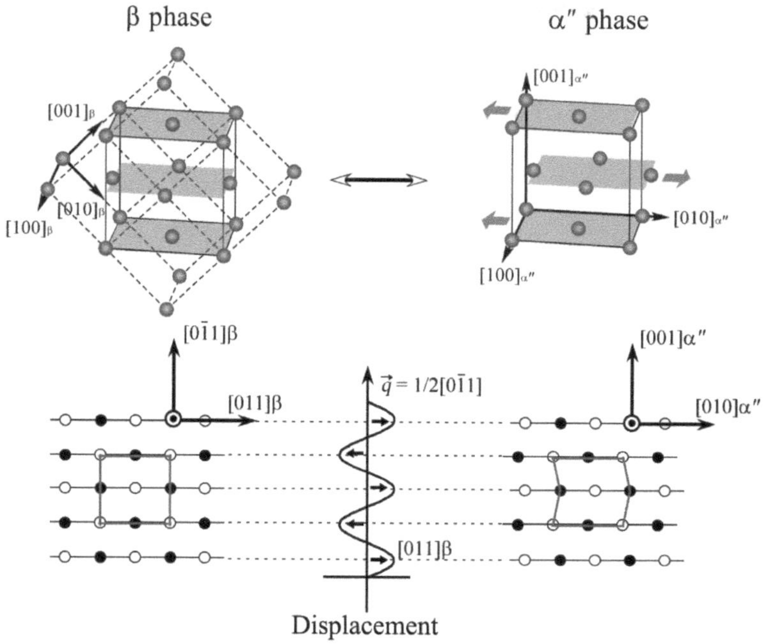

Fig. 3.15 A schematic illustration explaining the shuffling mode of the martensitic transformation from β to α″.

Table 3.1 Relationship between the shuffling modes and corresponding variant number in β-α″ martensitic transformation [23]

Corresponding variant number	Shuffling modes	Lattice correspondence		
		$[1\ 0\ 0]_{\alpha''}$	$[0\ 1\ 0]_{\alpha''}$	$[0\ 0\ 1]_{\alpha''}$
V1	$(0\bar{1}1)[011]_\beta$	$[100]_\beta$	$[011]_\beta$	$[0\bar{1}1]_\beta$
V2	$(011)[0\bar{1}1]_\beta$	$[\bar{1}00]_\beta$	$[0\bar{1}1]_\beta$	$[011]_\beta$
V3	$(10\bar{1})[101]_\beta$	$[010]_\beta$	$[101]_\beta$	$[10\bar{1}]_\beta$
V4	$(101)[10\bar{1}]_\beta$	$[0\bar{1}0]_\beta$	$[10\bar{1}]_\beta$	$[101]_\beta$
V5	$(\bar{1}10)[110]_\beta$	$[001]_\beta$	$[110]_\beta$	$[\bar{1}10]_\beta$
V6	$(110)[\bar{1}10]_\beta$	$[00\bar{1}]_\beta$	$[\bar{1}10]_\beta$	$[110]_\beta$

modulation modes are distributed equivalently in the β-phase matrix. As the oxygen content increases, the amount of nanodomains increases as shown in Fig. 3.17 [27].

Origin of lattice modulation and nanodomain structure is the local strain field induced by the interstitial alloying elements. It has been well acknowledged that oxygen atoms occupy the octahedral sites in the bcc lattice.

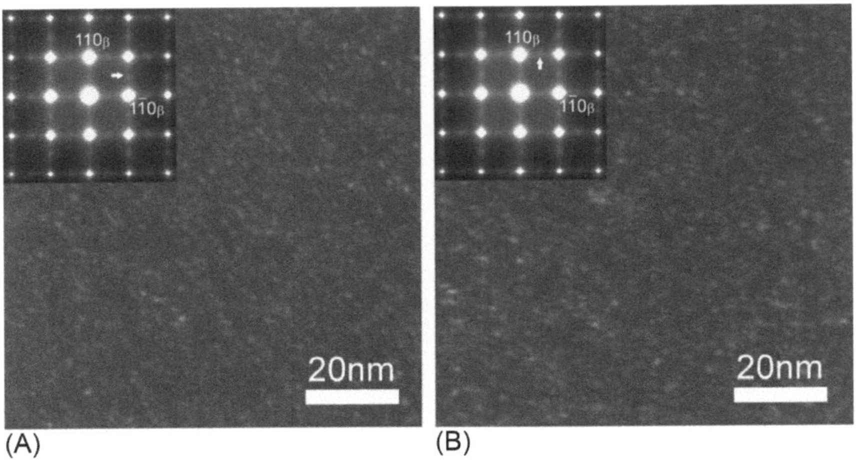

Fig. 3.16 Dark-field images of nanodomains in the β phase of a (Ti-23Nb)-1O alloy. (A) V6 $(110)\langle\bar{1}10\rangle$ and (B) V5 $(\bar{1}10)\langle 110\rangle$.

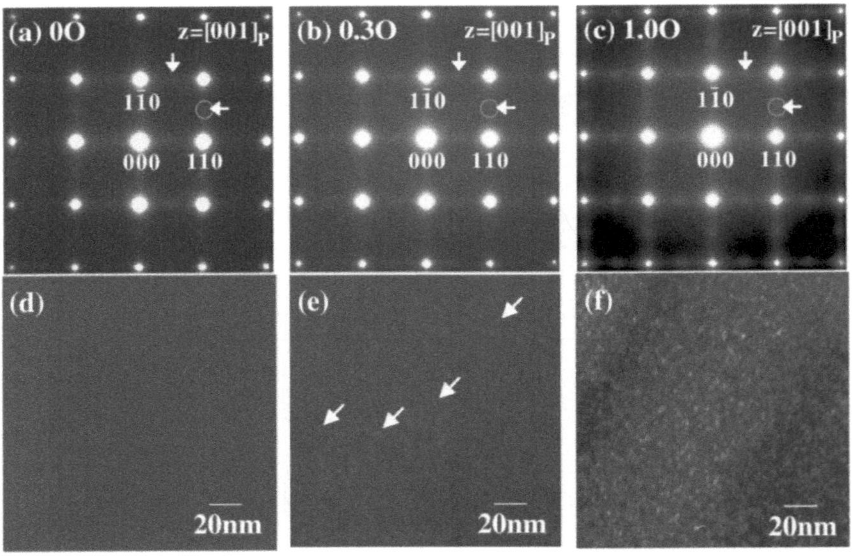

Fig. 3.17 Selected diffraction patterns of (A) Ti-26Nb, (B) Ti-26Nb-0.3O, and (C) Ti-26Nb-1.0O taken from [001] zone axis. (D), (E), and (F) are dark-field images of nanodomains from the superlattice spot surrounded by a circle in the diffraction pattern (A), (B), and (C), respectively.

A calculation using the embedded cell model also suggests that the interstitial oxygen atoms in the β phase of Ti–Nb alloys prefer to occupy the octahedral sites [42]. The octahedral sites in the bcc structure are classified into three groups as shown in Fig. 3.18A [43]. The octahedral site surrounded by six atoms is not symmetrical. As shown in Fig. 3.18B, in the case of A site, two atoms are in the [001] direction, and four atoms are in the [110] and $[1\overline{1}0]$ directions, respectively. Therefore, when the octahedral site is occupied by an interstitial atom, a local stress field with tetragonal symmetry is produced as shown in Fig. 3.18B: an interstitial atom that is located at A site induces the local stress field mainly along the [001] direction. The local stress field along the [001] direction can be relaxed by the shuffling of adjacent $(0\overline{1}1)$ planes in opposite directions, i.e., $[0\overline{1}\overline{1}]$ and [011], because the interatomic distance of nearest neighbor matrix atoms becomes larger, which is corresponding to the shuffling mode V1. The local stress field along the [001] direction can also be relaxed by the shuffling modes of $(011)[0\overline{1}1]$, $(10\overline{1})[101]$, and $(101)[10\overline{1}]$ planes, i.e., V2, V3, and V4 modes (Fig. 3.19 [22]). It is noted that these four shuffling modes are equivalent. Similarly, as shown in Fig. 3.19, the local stress field along the [100] direction can be relaxed by the shuffling modes V3, V4, V5, and V6. In the case of the local stress field along the [010], the shuffling modes V1, V2, V5, and V6 are favorable to relax the stress field. Similarly, the local stress field along the [001] can be relaxed by the shuffling modes V1, V2, V3, and V4. It is

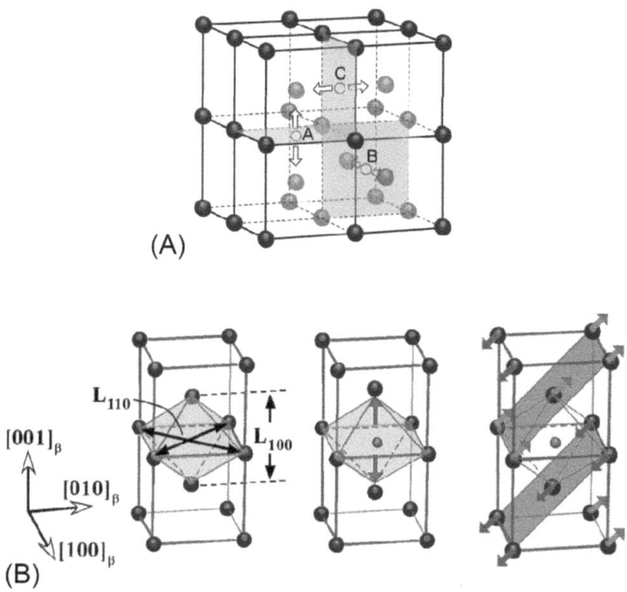

Fig. 3.18 A schematic illustration of (A) three type octahedral sites in the β phase and (B) the stress relaxation by the atomic shuffling.

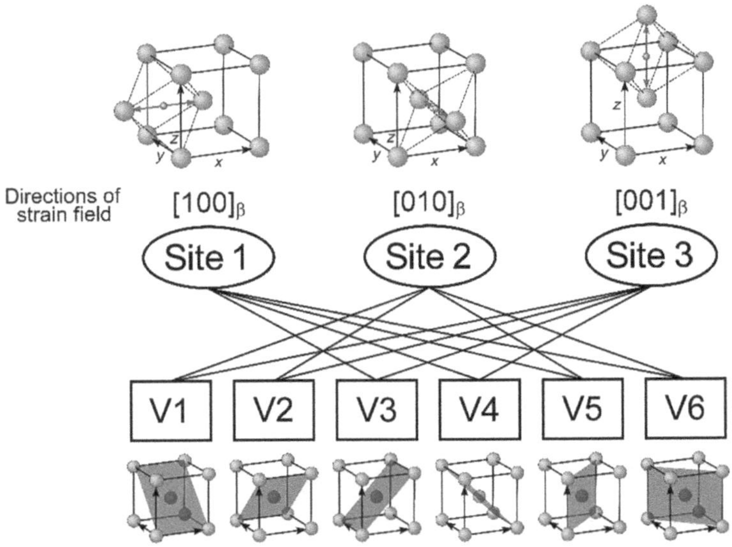

Fig. 3.19 A schematic illustration showing the relationship between the interstitial oxygen positions and shuffling mode.

reasonable to consider that, if the stress is not applied, three types of sites are equivalent and the interstitial atoms are distributed randomly and homogeneously. As a result, six shuffling modes are distributed equivalently in the matrix, resulting in that six variants of nanodomains are introduced equivalently. The formation of multiple variants of nanodomains hinders a long-range martensitic transformation because one nanodomain variant acts as a barrier preventing the growth of other nanodomain variants although the interstitial atoms facilitate the martensitic transformation locally.

3.5.2 Unusual Temperature Dependence

The nanodomain structure becomes unstable as temperature increases. Fig. 3.20 shows the change of selected diffraction pattern of the (Ti-23Nb)-1O alloy with temperature [22]. The strongest peak intensity of diffuse streaks is seen at 90 K. The peak intensity decreases with increasing temperature and becomes faint at 373 K and above, indicating that the nanodomain structure becomes unstable with increasing temperature and disappears at around 373 K. The stability of nanodomain structure is explained by the martensitic transformation temperature. The first-order martensitic transformation occurs in the Ti-23Nb alloy, where M_s and M_f are 378 and 313 K, respectively. The addition of oxygen induces the nanodomain structure and then

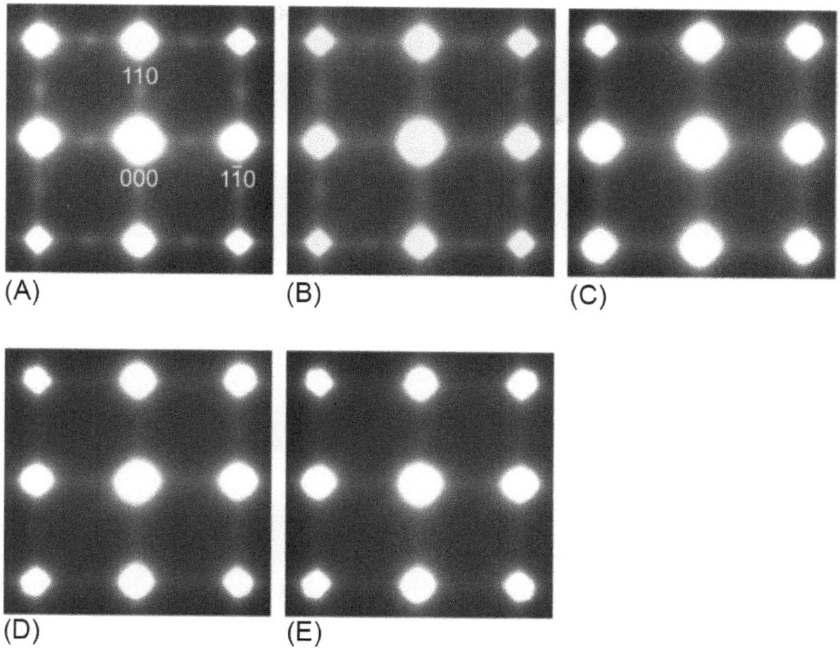

Fig. 3.20 Selected diffraction pattern of the (Ti-23Nb)-1O alloy taken from the [001] zone axis at (A) 90 K, (B) 293 K, (C) 353 K, (D) 373 K, and (E) 423 K.

suppresses the long-range martensitic transformation. The driving force for the martensitic transformation increases with decreasing temperature, resulting in more stable nanodomain structure at lower temperatures. On the other hand, with increasing temperature, the nanodomain structure becomes unstable not only because the β phase becomes more stable but also because the local strain field induced by the oxygen atom becomes weak due to thermal expansion of the β lattice.

Fig. 3.21 shows the temperature dependence of shape memory properties in the Ti-23Nb and (Ti-23Nb)-1O alloys [22]. The Ti-23Nb alloy exhibits shape memory effect in the temperature range 173–373 K; the specimens deformed in the temperature range 173–373 K recover their shape upon heating to 673 K. Partial superelastic recovery is seen at 423 K. At the temperatures higher than 673 K, the Ti-23Nb alloy exhibits only plastic deformation. On the other hand, the (Ti-23Nb)-1O alloy shows a quite different behavior. Superelastic recovery is observed at a temperature range of 173–323 K, while shape memory effect is observed at a higher temperature range of 323–423 K. This is opposite to the temperature dependence of shape memory

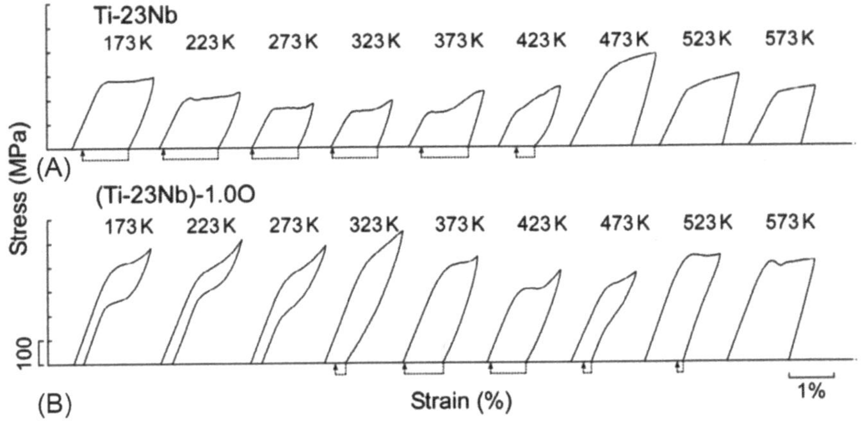

Fig. 3.21 Effect of deformation temperature on the stress-strain curve for (A) Ti-23Nb and (B) (Ti-23Nb)-1O alloys upon loading and unloading.

properties observed in ordinary shape memory alloys, unexplainable by thermodynamics of the martensitic transformation. It is also noted that the apparent yield stress for the superelastic deformation increases with decreasing temperature. These results indicate that the mechanism of superelastic recovery at temperature range of 173–323 K is different from ordinary shape memory alloys. It is suggested that superelastic behavior is due to the growth of some nanodomain variants, which are favorable to tensile direction. The lattice distortion of the favorable nanodomain variants increases continuously upon loading although the long-range martensitic transformation is suppressed by the presence of other variants. The local barrier of unfavorable nanodomain variants becomes stronger with decreasing temperature; therefore, the apparent yield stress increases as temperature decreases. The nanodomain structure becomes unstable, and the local barrier becomes weaker with increasing temperature. As a result, the long-range martensitic transformation can be induced upon loading at temperature range of 323–423 K. The stress-induced martensite phase is remained because A_s of the alloy is higher than 423 K. The reverse transformation of the remained martensite phase occurs by heating above A_f of the alloy, i.e., shape memory effect appears at temperatures higher than superelastic deformation.

3.5.3 Hysteresis Loop Rounding and Nonlinear Elasticity

Fig. 3.22 shows the effect of oxygen content on stress-strain curves of Ti-26Nb-xO alloys (x=0, 0.3, 0.5, 1.0) at different test temperatures

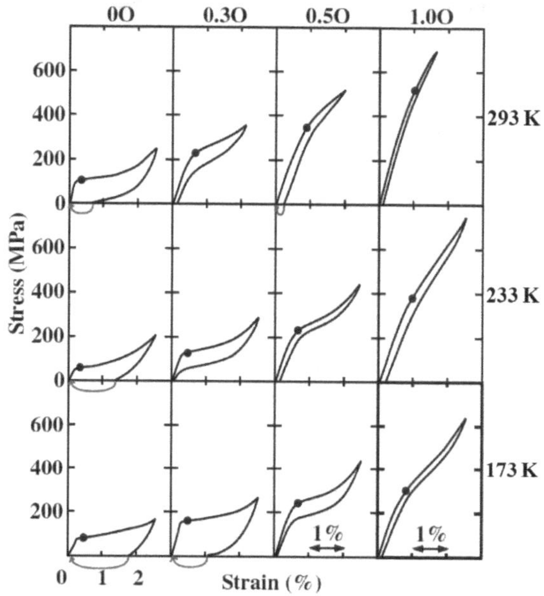

Fig. 3.22 Stress-strain curves upon loading and unloading for Ti-26Nb-(0–1)O alloys obtained at 173, 233, and 293 K.

[27]. The apparent yield stress in each stress-strain curve is indicated by a filled circle. It is clearly seen that the shape of stress-strain curves is strongly dependent on oxygen content. The Ti-26Nb alloy exhibits a stress plateau after the onset of the stress-induced martensitic transformation. After unloading, the residual strain is recovered by heating. The addition of oxygen in the Ti-26Nb causes an increase in the apparent yield stress and changes the transformation behavior, i.e., from shape memory effect to superelasticity. As the oxygen content increases, the stress plateau region disappears, and stress hysteresis becomes narrower progressively.

Fig. 3.23 compares the XRD profiles obtained by in situ XRD measurements during tensile deformation for the Ti-26Nb and Ti-26Nb-1O alloys [25]. XRD profiles were taken for the rolling plane, and the scattering vector was perpendicular to the applied stress. Prior to loading, i.e., at a strain of 0.0%, both Ti-26Nb and Ti-26Nb-1O alloys exhibit only β phase. The split of the 211_{β} peak is due to the peak doublet of Cu $K\alpha_1$ and Cu $K\alpha_2$ radiation. In the XRD profiles of the Ti-26Nb alloy, a reflection peak of $202_{\alpha''}$appears at $\varepsilon = 0.3\%$ where apparent yielding occurs. The intensity of the $202_{\alpha''}$peak increases at the expense of the 211_{β} peak with increasing strain in the stress plateau region, implying that the stress-induced martensitic transformation

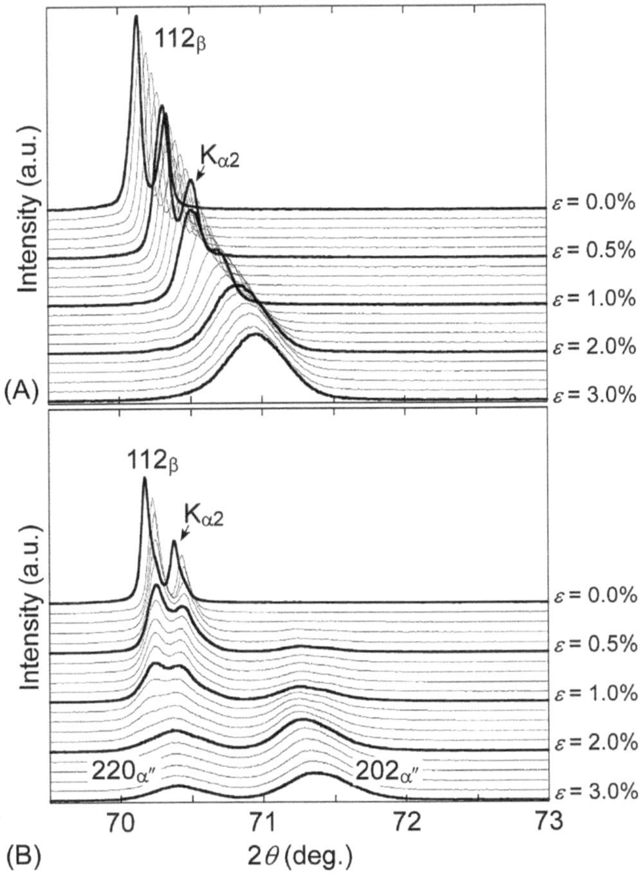

Fig. 3.23 XRD profiles obtained by in situ measurement for Ti-26Nb and Ti-26Nb-1O alloys.

proceeds in the Ti-26Nb alloy. On the other hand, the Ti-26Nb-1O alloy exhibits no distinct stress-induced martensitic transformation upon loading. The 211_β peak shifts toward a higher 2θ value continuously and becomes broader with increasing applied strain, indicating that the lattice distortion strain of the nanodomain increases with increasing applied stress.

Fig. 3.24 shows the strain-temperature curves under various constant stresses for Ti-26Nb-(0–1)O alloys [27]. Solid and broken lines indicate the strain change upon heating and cooling, respectively. The Ti-26Nb alloy exhibits strain-temperature loops with a large hysteresis owing to the martensitic transformation upon cooling and the reverse transformation upon heating, respectively. The transformation strain increases with increasing applied stress. It is noticed that the critical stress for the martensitic

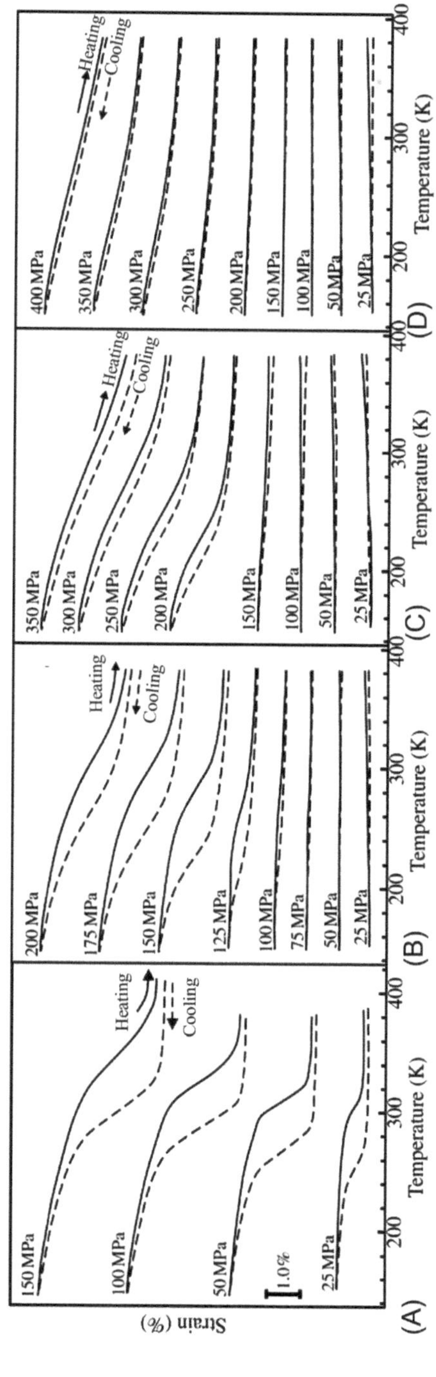

Fig. 3.24 Strain-temperature curves under various constant stresses for Ti-26Nb-(0–1)O alloys. (A) 0O, (B) 0.3O, (C) 0.5O, and (D) 1.0O.

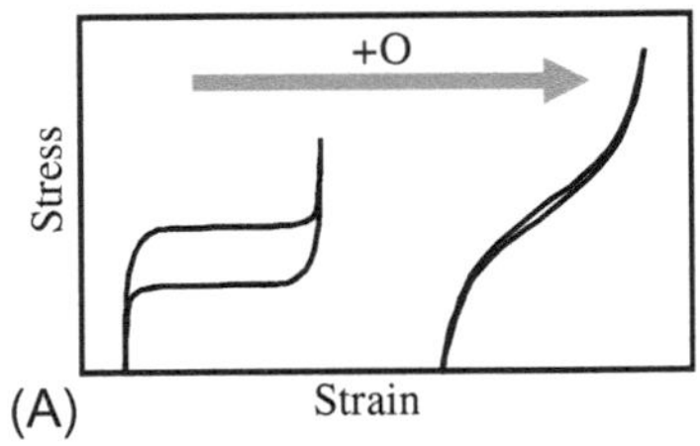

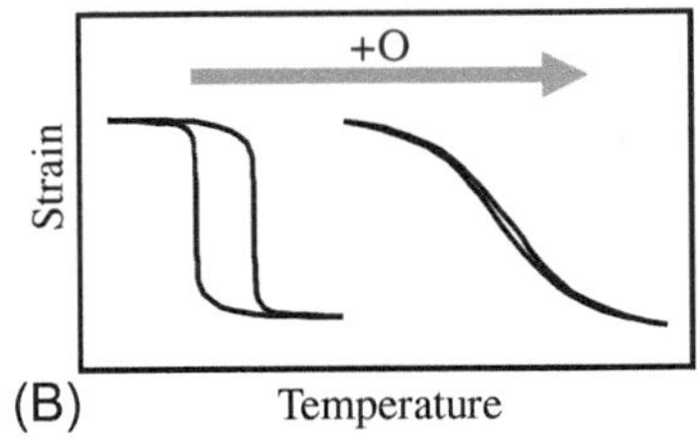

Fig. 3.25 Schematic illustrations explaining the effect of oxygen addition on (A) the stress-strain curve and (B) the stain-temperature curve.

transformation rises as the oxygen concentration increases. In addition hysteresis loop becomes round and slim with increasing oxygen concentration. The effect of interstitial alloying elements on the martensitic transformation behavior is shown schematically in Fig. 3.25. Both the effects of interstitial alloying elements on the stress-strain curve and on the strain-temperature curve can be explained by similar mechanisms. The interstitial atoms produce the local stress field and martensite like lattice modulation. Namely, the interstitial atoms facilitate the martensitic transformation locally and induce the nanodomain. However, the nanodomain structure hinders a long-range martensitic transformation because one nanodomain variant acts as a barrier preventing the growth of other nanodomain variants. Therefore, the stress for the growth of favorable variants of nanodomains or the increase in a lattice distortion increases continuously with increasing strain. As a result, the stress plateau disappears, and the stress hysteresis becomes small. Similar effect on the strain-temperature curve under a constant stress is explained as follows. In an alloy with a few interstitial alloying elements, when the martensitic transformation occurs at a certain temperature (M_s) upon cooling, the transformation progresses in a narrow temperature range, causing a sharp increase of strain and distinct hysteresis. The increase in the number of interstitial alloying elements suppresses the progress of the martensitic transformation due to the pinning effect of nanodomains. Therefore, in order to progress the martensitic transformation, further cooling is required, leading to a slim and round strain-temperature hysteresis.

3.6 SUMMARY

The interstitial alloying elements, particularly O and N, significantly affect not only the martensitic transformation behavior of Ti-based alloys but also their shape memory properties. The addition of O or N improves the superelastic properties of Ti-Nb alloys by increasing the critical stress for slip. Furthermore, the addition of O and N reduces apparent martensitic transformation temperature of Ti-Nb alloys by 160–200 K per 1 at.%, resulting in a larger transformation strain at superelastic compositions with less Nb content. The interstitial atoms in the metastable β phase produce a nanosized $\{110\}\langle 1\bar{1}0\rangle$-type lattice modulation (nanodomain) structure to relax local strain fields of interstitial alloying elements, where the transverse displacement is crystallographically equivalent to the shuffling of basal plane of α'' martensite phase. The suppression of long-range martensitic transformation and the decrease of the apparent martensitic transformation temperature are due to the formation of multiple variants of nanodomains. As the concentration of interstitial elements increases, hysteresis loops of stress-strain curves and strain-temperature curves become round and slim. The nonlinear elasticity in the Ti-Nb alloys containing higher amount of oxygen can be explained by the nanodomain structure. On the other hand, C and B have a relatively weaker effect on the martensitic transformation behavior and superelastic properties of Ti-Nb-based alloys due to their limited solubility in the β phase.

REFERENCES

[1] H. Conrad, Effect of interstitial solutes on the strength and ductility of titanium, Prog. Mater. Sci. 26 (1981) 123–403.

[2] M.L. Wasz, F.R. Brotzebm, R.B. McLellan, A.J. Griffin, Effect of oxygen and hydrogen on mechanical properties of commercial purity titanium, Int. Mater. Rev. 41 (1996) 1–12.

[3] M. Yan, W. Xu, M.S. Dargusch, H.P. Tang, M. Brandt, M. Qian, Review of effect of oxygen on room temperature ductility of titanium and titanium alloys, Powder Metall. 57 (2014) 251–257.

[4] Q. Yu, L. Qi, T. Tsuru, R. Traylor, D. Rugg, J.W. Morris Jr, M. Asta, D.C. Chrzan, A. M. Minor, Origin of dramatic oxygen solute strengthening effect in titanium, Science 347 (2015) 635–639.

[5] J.C. Williams, A.W. Somer, P.P. Tung, The influence of oxygen concentration on the internal stress and dislocation arrangements in α titanium, Metall. Trans. 3 (1972) 2979–2984.

[6] R. Boyer, G. Welsch, E.W. Collings, Materials Properties Handbook: Titanium Alloys, ASM, Materials Park, OH, 1994.

[7] M.C. Brandes, M. Baughman, M.J. Mills, J.C. Williams, The effect of oxygen and stress state on the yield behavior of commercially pure titanium, Mater. Sci. Eng. A 551 (2012) 13–18.
[8] F. Geng, M. Niinomi, M. Nakai, Observation of yielding and strain hardening in a titanium alloy having high oxygen content, Mater. Sci. Eng. A 551 (2011) 5435–5445.
[9] E.G. Obbard, Y.L. Hao, R.J. Talling, S.J. Li, Y.W. Zhang, D. Dye, R. Yang, The effect of oxygen on α'' martensite and superelasticity in Ti–24Nb–4Zr–8Sn, Acta Mater. 59 (2011) 112–125.
[10] M. Tane, T. Nakano, S. Kuramoto, M. Hara, M. Niinomi, N. Takesue, T. Yano, H. Nakajima, Low Young's modulus in Ti–Nb–Ta–Zr–O alloys: cold working and oxygen effects, Acta Mater. 59 (2011) 6975–6988.
[11] M. Besse, P. Castany, T. Gloriant, Mechanisms of deformation in gum metal TNTZ-O and TNTZ titanium alloys: a comparative study on the oxygen influence, Acta Mater. 59 (2011) 5982–5988.
[12] J.I. Kim, H.Y. Kim, H. Hosoda, S. Miyazaki, Shape memory behavior of Ti–22Nb–(0.5–2.0)O (at%) biomedical alloys, Mater. Trans. 46 (2005) 852–857.
[13] A. Ramarolahy, P. Castany, F. Prima, P. Laheurte, I. Peron, T. Gloriant, Microstructure and mechanical behavior of superelastic Ti–24Nb–0.5O and Ti–24Nb–0.5N biomedical alloys, J. Mech. Behav. Biomed. Mater. 9 (2012) 83–90.
[14] T. Saito, T. Furuta, J.H. Hwang, S. Kuramoto, K. Nishino, N. Suzuki, et al., Multifunctional alloys obtained via a dislocation-free plastic deformation mechanism, Science 300 (2003) 464–467.
[15] R.J. Talling, R.J. Dashwood, M. Jackson, D. Dye, On the mechanism of superelasticity in gum metal, Acta Mater. 57 (2009) 1188–1198.
[16] Y. Yang, S.Q. Wu, G.P. Li, Y.L. Li, Y.F. Lu, K. Yang, P. Ge, Evolution of deformation mechanisms of Ti–22.4Nb–0.73Ta–2Zr–1.34O alloy during straining, Acta Mater. 58 (2010) 2778–2787.
[17] S. Kuramoto, T. Furuta, J. Hwang, K. Nishino, T. Saito, Elastic properties of gum metal, Mater. Sci. Eng. A 442 (2006) 454–457.
[18] M.Y. Gutkin, T. Ishizaki, S. Kuramoto, I.A. Ovid'ko, Nanodisturbances in deformed gum metal, Acta Mater. 54 (2006) 2489–2499.
[19] H. Xing, J. Sun, Q. Yao, W.Y. Guo, R. Chen, Origin of substantial plastic deformation in gum metals, Appl. Phys. Lett. 92 (2008)151905.
[20] J.W. Morris, Y. Hanlumyuang, M. Sherburne, E. Withey, D.C. Chrzan, S. Kuramoto, Y. Hayashi, Anomalous transformation-induced deformation in <110> textured gum metal, Acta Mater. 58 (2010) 3271–3280.
[21] F. Yin, S. Iwasaki, D. Ping, K. Nagai, Sneok-type high-damping alloys realized in β-Ti alloys with high oxygen solid solution, Adv. Mater. 18 (2006) 1541–1544.
[22] M. Tahara, T. Kanaya, H.Y. Kim, T. Inamura, H. Hosoda, S. Miyazaki, Heating-induced martensitic transformation and time-dependent shape memory behavior of Ti–Nb–O alloy, Acta Mater. 80 (2014) 317–326.
[23] H.Y. Kim, L.S. Wei, S. Kobayashi, M. Tahara, S. Miyazaki, Nanodomain structure and its effect on abnormal thermal expansion behavior of a Ti–23Nb–2Zr–0.7Ta–1.2O alloy, Acta Mater. 67 (2013) 4874–4886.
[24] L.S. Wei, H.Y. Kim, S. Miyazaki, Effects of oxygen concentration and phase stability on nano-domain structure and thermal expansion behavior of Ti–Nb–Zr–Ta–O alloys, Acta Mater. 100 (2015) 313–322.
[25] M. Tahara, H.Y. Kim, T. Inamura, H. Hosoda, S. Miyazaki, Role of interstitial atoms in the microstructure and non-linear elastic deformation behavior of Ti–Nb alloy, J. Alloys Compd. 577S (2013) S404–S407.
[26] M. Tahara, H.Y. Kim, T. Inamura, H. Hosoda, S. Miyazaki, Lattice modulation and superelasticity in oxygen-added β-Ti alloys, Acta Mater. 59 (2011) 6208–6218.

[27] Y. Nii, T. Arima, H.Y. Kim, S. Miyazaki, Effect of randomness on ferroelastic transitions: disorder-induced hysteresis loop rounding in Ti–Nb–O martensitic alloy, Phys. Rev. B 82 (2010) 214104.
[28] H.Y. Kim, S. Miyazaki, Several issues in the development of Ti–Nb-based shape memory alloys, Shape Mem. Superelasticity 2 (2016) 380–390.
[29] M. Tahara, H.Y. Kim, H. Hosoda, S. Miyazaki, Shape memory effect and cyclic deformation behavior of Ti–Nb–N alloys, Funct. Mater. Lett. 2 (2009) 79–82.
[30] M. Tahara, H.Y. Kim, H. Hosoda, T.H. Nam, S. Miyazaki, Effect of nitrogen addition and annealing temperature on superelastic properties of Ti–Nb–Zr–Ta alloys, Mater. Sci. Eng. A 527 (2010) 6844–6852.
[31] M. Tahara, H.Y. Kim, T. Inamura, H. Hosoda, S. Miyazaki, Effect of nitrogen addition on superelasticity of Ti–Zr–Nb alloys, Mater. Trans. 50 (2009) 2726–2730.
[32] Y. Al-Zain, H.Y. Kim, S. Miyazaki, Effect of B addition on the microstructure and superelastic properties of a Ti–26Nb alloy, Mater. Sci. Eng. A 644 (2015) 85–89.
[33] Y. Horiuchi, T. Inamura, H. Hosoda, K. Wakashima, H.Y. Kim, S. Miyazaki, Effect of boron addition on transformation behavior and tensile properties of Ti–Nb–Al alloy, Mater. Sci. Eng. A 438–440 (2006) 830–834.
[34] Y. Horiuchi, T. Inamura, H.Y. Kim, K. Wakashima, S. Miyazaki, H. Hosoda, Effect of boron concentration on martensitic transformation temperatures, stress for inducing martensite and slip stress of Ti–24 Mol%Nb–3mol% Al superelastic alloy, Mater. Trans. 48 (2007) 407–413.
[35] H. Hosoda, Y. Horiuchi, T. Inamura, K. Wakashima, H.Y. Kim, S. Miyazaki, Effect of carbon addition of shape memory properties of TiNb alloys, Mater. Sci. Forum 638–642 (2010) 2046–2051.
[36] G.D. Sandrock, A.J. Perkins, R.F. Hehemann, The premartensitic instability in near-equiatomic TiNi, Metall. Mater. Trans. B Process Metall. Mater. Process. Sci. 2 (1971) 2769–2781.
[37] S. Sarkar, X. Ren, K. Otsuka, Evidence for strain glass in the ferroelastic-martensitic system $Ti_{50-x}Ni_{50+x}$, Phys. Rev. Lett. 95 (2005) 205702.
[38] Y. Murakami, D. Shindo, Lattice modulation preceding to the R-phase transformation in a Ti50Ni48Fe2 alloy studied by TEM with energy-filtering, Mater. Trans. 40 (1999) 1092–1097.
[39] M. Todai, T. Fukuda, T. Kakeshita, Relation between incommensurate satellites and phonon softening in Ti–Ni-based shape memory alloys, Scr. Mater. 64 (2011) 541–543.
[40] K. Otsuka, C.M. Wayman, H. Kubo, Diffuse electron scattering in β-phase alloys, Metall. Mater. Trans. A 9 (1978) 1075–1085.
[41] K. Enami, K.J. Hasunuma, A. Nagasawa, S. Nenno, Elastic softening and electron-diffraction anomalies prior to the martensitic transformation in a Ni–Al β_1 alloy, Scr. Metall. 10 (1976) 879–884.
[42] L. Yu, F. Yin, D. Ping, Natural mechanism of the broadened Snoek relaxation profile in ternary body-centered-cubic alloys, Phys. Rev. B 75 (2007) 174105.
[43] H.Y. Kim, S. Miyazaki, Martensitic transformation and superelastic properties of Ti–Nb base alloys, Mater. Trans. 56 (2015) 625–634.

CHAPTER 4

Thermomechanical Treatment and Microstructure Control

Contents

4.1 INTRODUCTION

A high resistance to plastic deformation is an indispensable property for reliable operation of the shape memory and superelastic functions. In Ti-Ni shape memory alloys, two strengthening mechanisms, namely, precipitation hardening and work hardening, have been mainly utilized to improve functionality [1]. Particularly, the formation of fine Ti_3Ni_4 precipitates has been reported to be very effective in increasing the critical stress for plastic deformation [2, 3]. Enhancement of strength is one of the key issues in β-type Ti-based shape memory alloys because of the low critical stress for slip originating from the disordered bcc structure [4–6]. There have been extensive efforts to improve the strength of the β-type Ti-based shape memory alloys. It has been confirmed that the major strengthening mechanisms, i.e., (1) solid-solution hardening, (2) work hardening, (3) grain size refinement, and (4) precipitation hardening, are effective to increase the critical stress for plastic deformation of the β-type Ti-based shape memory alloys

Ni-free Ti-based Shape Memory Alloys
https://doi.org/10.1016/B978-0-12-809401-3.00004-5

[7–23]. As described in Chapter 3, among the alloying elements, interstitial elements such as O and N have most pronounced solid-solution hardening effect. The latter three methods are associated with the microstructure control through thermomechanical treatment. Texture is another important microstructural parameter to affect the shape memory and superelastic properties. In this chapter, the effects of thermomechanical treatment on microstructure, texture, and shape memory and superelastic properties of β-type Ti-based shape memory alloys are summarized. Relationships between microstructure parameters and shape memory properties are discussed.

4.2 ANNEALING

4.2.1 Effect of Annealing Temperature

Fig. 4.1 shows the effect of annealing temperature, where the annealing time was 3.6 ks, on the stress-strain curves of Ti-22Nb-6Zr specimens cold-rolled with a reduction of 95% in thickness [13]. The specimen annealed at the lowest temperature of 773 K exhibited highest strength and lowest ductility, which is due to not only the low annealing temperature which is too low for the recovery of highly deformed structure but also the precipitation of α phase. All specimens except annealed at 773 K show double-yielding behavior, indicating that the stress-induced martensitic transformation occurs

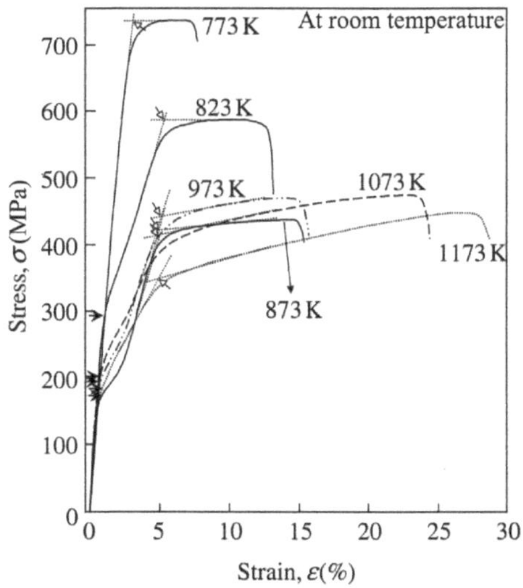

Fig. 4.1 Stress-strain curves of the Ti-22Nb-6Zr specimens annealed at different temperatures between 773 and 1173 K for 3.6 ks after cold rolling up to 95%.

prior to the plastic deformation in the specimens annealed at temperatures between 823 and 1173K. The second yielding point, i.e., the critical stress for plastic deformation, tends to decrease with increasing annealing temperature; however, the fracture strain shows the opposite dependence on the annealing temperature. The onset stress for the stress-induced martensitic transformation also decreases with increasing annealing temperature particularly in the temperature range between 823 and 873K.

Fig. 4.2 shows TEM images of the specimens annealed at 823 and 873K for 3.6ks [13]. The specimen annealed at 823K exhibits a subgrain structure with an average grain size of 1.5 μm. In addition, fine secondary α precipitates are seen in the specimen annealed at 823K. On the other hand, no α precipitates are seen in the specimen annealed at 873K, indicating that the β transus temperature of the Ti-22Nb-6Zr alloy is between 823 and 873K. It is also seen that the specimen annealed at 873K shows a bimodal structure composed of fine subgrains and coarse recrystallized grains, implying that the recrystallization starts to occur at 873K.

The superelastic properties of the specimens annealed at 823, 873, and 973K are compared in Fig. 4.3 [13]. It is noted that the specimen annealed at 823K shows a good superelasticity with a narrow stress hysteresis when compared with the specimens annealed at higher temperatures. Such good superelasticity of the specimen annealed at 823K, in spite of a higher onset stress for the stress-induced martensitic transformation, originates from the higher critical stress for plastic deformation due to the fine subgrain structure and α precipitates.

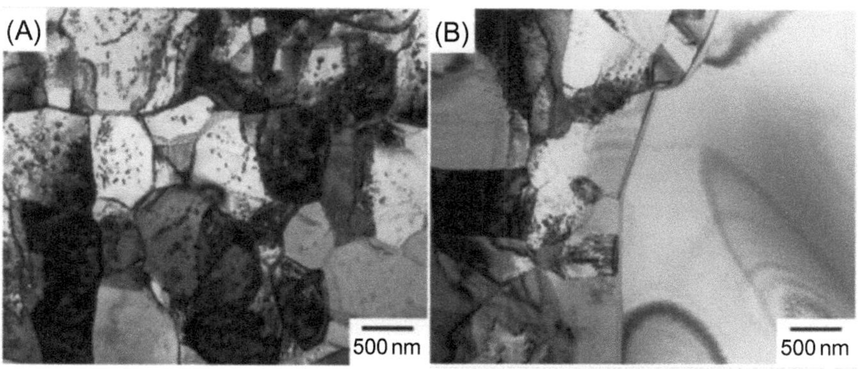

Fig. 4.2 TEM bright-field images of the Ti-22Nb-6Zr specimens annealed at (A) 823 and (B) 873K.

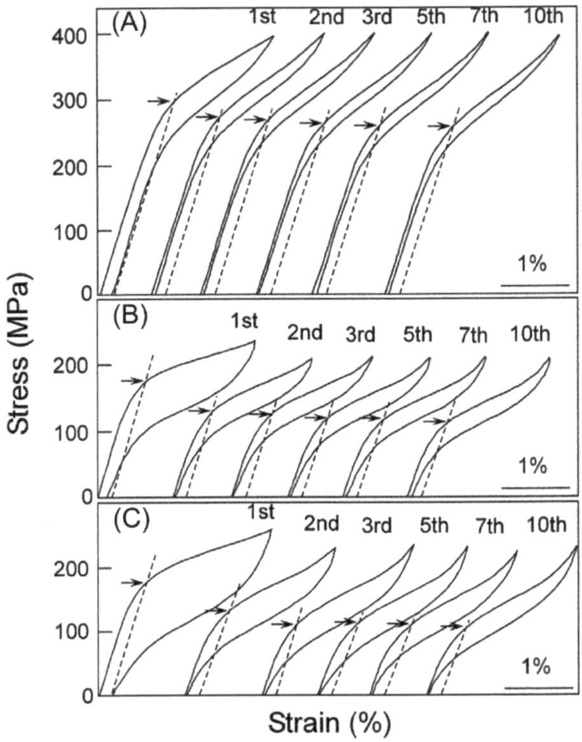

Fig. 4.3 Stress-strain curves obtained by loading and unloading tensile tests of Ti-22Nb-6Zr specimens annealed at (A) 823, (B) 873, and (C) 973 K.

4.2.2 Effect of Composition on Optimum Annealing Condition

The optimum annealing temperature is strongly dependent on the composition of alloy because not only recrystallization temperature but also β transus temperature is strongly affected by alloy composition. Fig. 4.4 shows bright-field TEM micrographs of Ti-22Nb-4Zr-2Ta and Ti-20Nb-4Zr-2Ta-0.6N alloys annealed at temperatures between 773 and 1073 K for 3.6 ks after cold rolling with a reduction of 98.5% in thickness [10]. The Ti-22Nb-4Zr-2Ta specimen annealed at 773 K shows a fine subgrain structure with a rather high dislocation density in the subgrains. The presence of α phase was confirmed in the selected area diffraction pattern of the Ti-22Nb-4Zr-2Ta specimen annealed at 773 K. Well-developed subgrain structure with a low dislocation density in the subgrains was found in the Ti-22Nb-4Zr-2Ta specimen annealed at 873 K. The α phase was observed mainly at grain boundaries of the β phase. Grain size gradually increases with increasing annealing

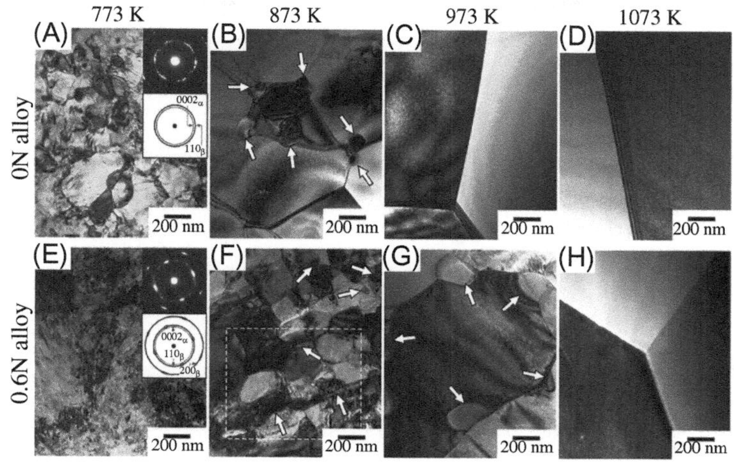

Fig. 4.4 TEM micrographs of the (A)–(D) Ti-22Nb-4Zr-2Ta and (E)–(H) Ti-20Nb-4Zr-2Ta-0.6N alloys annealed at 773–1073 K for 3.6 ks. *White arrows* in B, F, and G indicate the α-phase precipitates.

temperature. Only the β phase was observed in the Ti-22Nb-4Zr-2Ta specimens annealed at 973 and 1073 K, indicating that the β transus temperature of the Ti-22Nb-4Zr-2Ta is between 873 and 973 K.

Similarly to the Ti-22Nb-4Zr-2Ta specimen, the recovery and recrystallization proceeded as the annealing temperature increases in the Ti-20Nb-4Zr-2Ta-0.6N specimen. However, it is clear that the recovery and recrystallization were relatively retarded in Ti-20Nb-4Zr-2Ta-0.6N specimen compared with Ti-22Nb-4Zr-2Ta. The Ti-20Nb-4Zr-2Ta-0.6N specimen annealed at 773 K exhibited a highly deformed structure, while the Ti-20Nb-4Zr-2Ta specimen annealed at the same temperature showed a subgrain structure. Also, the specimens annealed at 873 K exhibited different microstructures in both alloys: much higher dislocation density and smaller grain size were observed in the Ti-20Nb-4Zr-2Ta-0.6N specimen. Furthermore, the precipitation of the α phase was confirmed up to 973 K in the Ti-20Nb-4Zr-2Ta-0.6N specimen, indicating that the β transus temperature of the Ti-22Nb-4Zr-2Ta-0.6N is between 973 and 1073 K. These results are consistent with the fact that N is a strong α-phase stabilizer and increases the β transus temperature. This result plausibly confirms that the precipitation of α precipitates retards the recovery and recrystallization of the β phase.

Fig. 4.5 compares stress-strain curves obtained in the specimens annealed at temperatures between 773 and 1173 K for Ti-22Nb-4Zr-2Ta and Ti-20Nb-4Zr-2Ta-0.6N alloys [10]. In both alloys, the specimens annealed

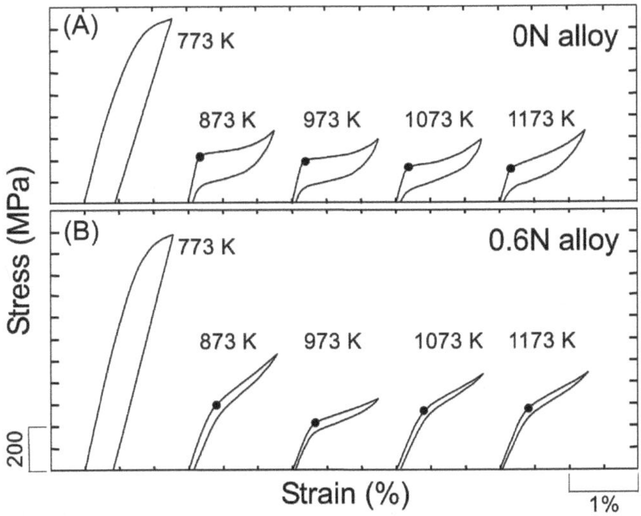

Fig. 4.5 Stress-strain curves of the (A) Ti-22Nb-4Zr-2Ta and (B) Ti-20Nb-4Zr-2Ta-0.6N alloys annealed at 773–1073 K for 3.6 ks. The *solid circle* indicates the apparent yield stress that corresponds to the onset stress for inducing martensitic transformation.

at 773 K exhibited a high yield strength, but no superelastic recovery was observed. This is because the fine subgrain structure and a high dislocation density in the specimens annealed at low temperature of 773 K, as shown in Fig. 4.4A and E, suppressed the stress-induced martensitic transformation. Other specimens annealed at higher temperatures exhibited good superelasticity. In the case of the Ti-20Nb-4Zr-2Ta-0.6N alloy, the onset stress for the stress-induced martensitic transformation, indicated by a closed circle, decreased with increasing annealing temperature, which is consistent with other alloys such as Ti-22Nb-6Zr as shown in Fig. 4.3. The decrease in the onset stress for the stress-induced martensitic transformation is associated with the increase of the martensitic transformation temperature that is due to the microstructure change. Not only the decrease in dislocation density but also the increase in grain size causes the increase in the martensitic transformation temperature, leading to the decrease in the onset stress of the stress-induced martensitic transformation. Also, the dissolution of α phase acts to facilitate the martensitic transformation because the Ti content of the matrix increases, causing the increase in the martensitic transformation temperature.

On the other hand, in the Ti-20Nb-4Zr-2Ta-0.6N alloy, the onset stress for the stress-induced martensitic transformation did not change monotonically with annealing temperature. The onset stress decreased with increasing annealing temperature to a minimum value at 973 K; then, it increased by further increasing annealing temperature. The decrease in the onset stress with increasing annealing temperature from 773 to 973 K is explained by the microstructure change such as the decrease in the dislocation density and the progress of recovery as shown in Fig. 4.4, as in the case of the Ti-22Nb-4Zr-2Ta. However, the microstructure change could not explain the increase of the onset stress with further increasing annealing temperature from 973 K. It was suggested that this can be explained by the scavenging effect of the α phase. It has been well documented that the α phase acts as a scavenger of interstitial alloying elements such as oxygen and nitrogen. Fig. 4.6 shows a schematic illustration of the composition change in the β and α phases when the specimen is heat treated at above and below the β transus temperature [10]. If the specimen is heat treated at below the β transus temperature, the Ti content in the β phase becomes less because the Ti is more enriched in the α phase. This change causes an increase in the martensitic transformation temperature of the alloy, resulting in the decrease in the onset stress for the stress-induced martensitic transformation with increasing annealing temperature between 873 and 973 K in the Ti-22Nb-4Zr-2Ta alloys. The content of N in the β phase is strongly dependent on the presence of the α phase. When the Ti-20Nb-4Zr-2Ta-0.6N alloy is heat treated at below the β transus temperature, N is preferentially partitioned into the α phase. Therefore, the dissolution of the α phase by the heat treatment at above the β transus temperature results in the increase in the N content in the β phase and the decrease in the martensitic

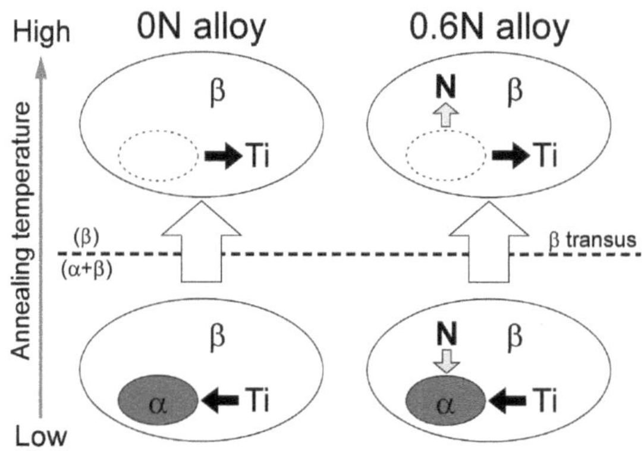

Fig. 4.6 A schematic illustration explaining the composition change due to the dissolution of the α phase in the Ti-22Nb-4Zr-2Ta and Ti-20Nb-4Zr-2Ta-0.6N alloys.

transformation temperature. As a result, in the Ti-20Nb-4Zr-2Ta-0.6N alloy, the onset stress increases with increasing annealing temperature across the β transus temperature, although the dissolution of Ti into the β phase increases the martensitic transformation temperature, because the effect of the N content in the martensitic transformation temperature is much stronger than the effect of the Ti content. Optimal annealing temperature is dependent on the composition of the alloys. Fig. 4.7 shows the annealing temperature dependence of the critical stress for plastic deformation and superelastic recovery strain for the Ti-22Nb-4Zr-2Ta and Ti-20Nb-4Zr-2Ta-0.6N alloys [10]. For the Ti-22Nb-4Zr-2Ta alloy, the critical stress for plastic deformation decreases with increasing annealing temperature monotonically due to the recovery and recrystallization as mentioned above. The decrease in the critical stress for plastic deformation causes the decrease in the superelastic recovery strain. For the Ti-20Nb-4Zr-2Ta-0.6N, both the critical stress for the plastic deformation and the recovery strain increase with increasing annealing temperature that is due to the solution strengthening effect of N in the β phase as mentioned above.

4.3 AGING

4.3.1 Effect of Aging Temperature and Time

Fig. 4.8 shows the effect of aging temperature on stress-strain curves of a Ti-26Nb alloy [23]. All specimens show a distinct double-yielding behavior.

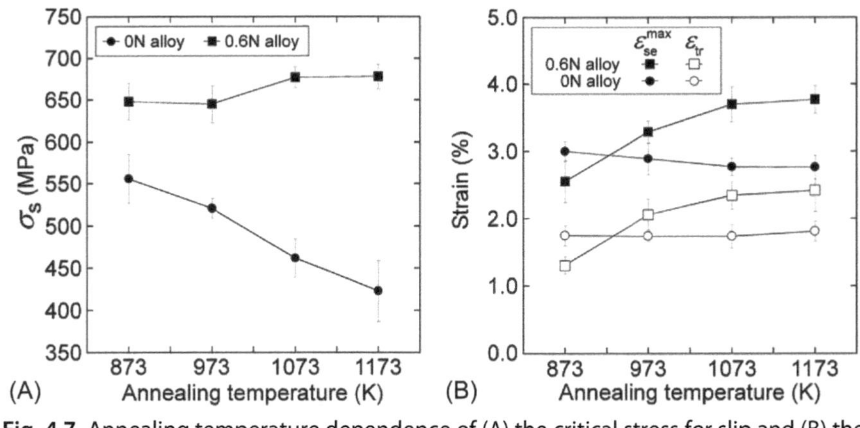

Fig. 4.7 Annealing temperature dependence of (A) the critical stress for slip and (B) the maximum superelastic recovery strain and transformation strain for the Ti-22Nb-4Zr-2Ta and Ti-20Nb-4Zr-2Ta-0.6N alloys.

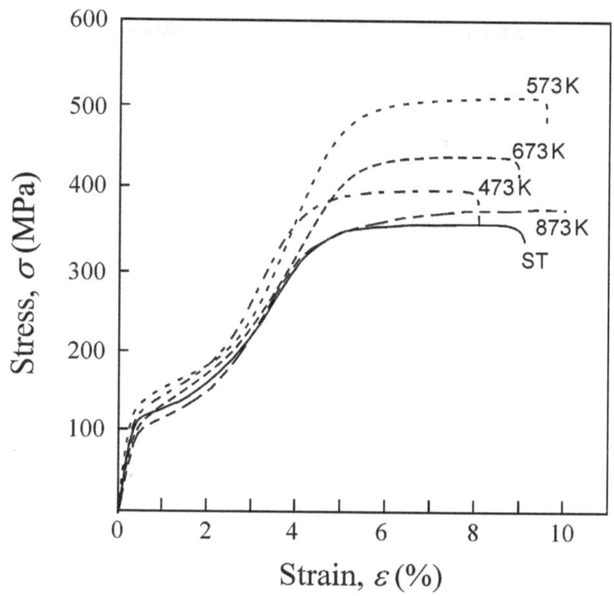

Fig. 4.8 Stress-strain curves of the Ti-26Nb alloy aged at different temperatures between 473 and 873 K for 3.6 ks after solution treatment at 1173 K for 1.8 ks.

The stress at which second yielding occurs increases by aging treatment, indicating that the critical stress for plastic deformation of stress-induced martensite phase increases by aging. The strengthening effect becomes stronger as the aging temperature increases up to 573 K and then decreases by further increasing aging temperature. The specimen aged at 873 K shows no strengthening effect. The strengthening effect by aging in the Ti-26Nb alloy is due to the formation of the ω phase that is called thermal ω phase. Fig. 4.9 shows the dark-field TEM images and selected area diffraction patterns of the aged specimens and solution-treated specimen of the Ti-26Nb alloy [23]. The dark-field images were taken from the diffraction spot of the ω phase. Diffuse scattering due to the ω phase is seen in the SADP of the solution-treated specimen. Very fine ω particles were observed in the dark-field image of the solution-treated specimen. This ω phase, so-called athermal ω phase, was formed upon quenching after the solution treatment. With increasing aging temperature up to 573 K, the peak intensity of the diffraction spots of the ω phase became strong, and the size and volume fraction of ω particles in the dark-field image increased remarkably. Further increase in aging temperature from 573 to 673 K resulted in the decrease

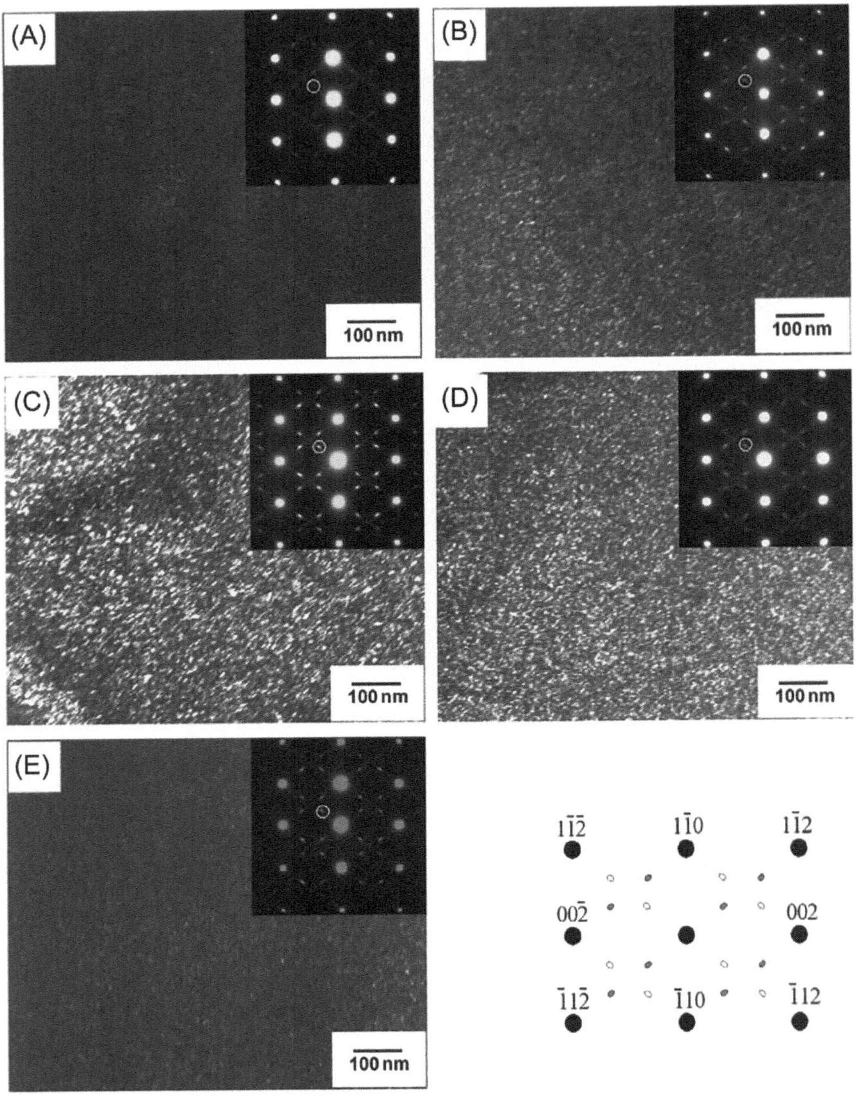

Fig. 4.9 Dark-field TEM micrographs and the corresponding selected area diffraction patterns of (A) the solution-treated Ti-26Nb alloy and specimens aged at (B) 473, (C) 573, (D) 673, and (E) 873 K for 3.6 ks.

of the ω phase. The specimen aged at 873 K exhibited a similar microstructure to that of the solution-treated specimen. Therefore, it is very clear that the strengthening effect by aging is directly caused by the precipitation of the ω phase.

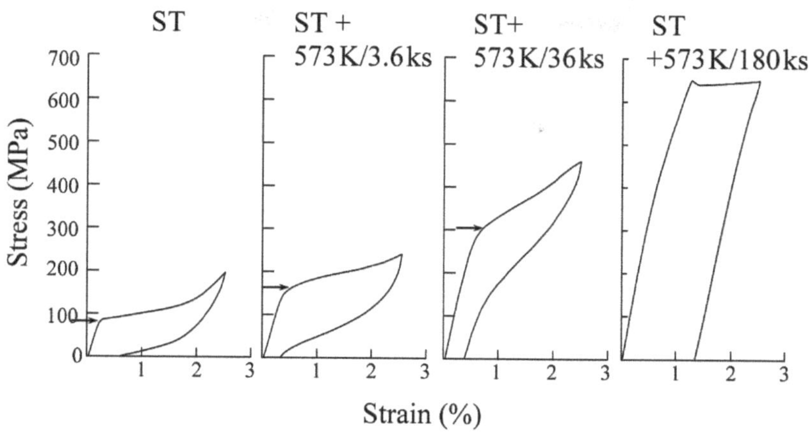

Fig. 4.10 Stress-strain curves obtained upon loading and unloading for the Ti-26Nb specimen solution treated at 1173 K and the specimens aged at 573 K for 3.6, 36, and 180 ks.

Fig. 4.10 shows the effect of aging at 573 K on superelastic properties of the Ti-26Nb alloy solution treated at 1173 K [24]. The solution-treated specimen exhibited superelastic recovery upon unloading. The onset stress for the stress-induced martensitic transformation increased with increasing aging time up to 36 ks. The onset stress for the stress-induced martensitic transformation increased with increasing aging time, implying that the martensitic transformation temperature decreased by aging. The increase in the onset stress for the stress-induced martensitic transformation generally leads to deterioration of superelastic properties because the permanent deformation is introduced more easily at a higher stress. However, the superelastic recovery was improved by aging up to 36 ks. This is due to the fact that the strengthening effect by the precipitation of the ω phase is stronger than the effect of the decrease in the transformation temperature. On the other hand, the longer aging time of 180 ks resulted in poor superelastic properties due to a high volume fraction of the ω phase, which decreased the transformation temperature too much thus increasing the onset stress for the stress induced martensitic transformation higher than the stress for plastic deformation.

Fig. 4.11 shows the effect of aging at 573 K on the superelastic properties of a Ti-26Nb alloy annealed at 873 K for 0.6 ks after cold rolling [25]. Similarly to the solution-treated specimen, the superelastic properties are improved by aging. In particular, the specimen aged for 3.6 ks exhibited excellent superelastic properties up to the fifth cycle, which is due to the combined effect of work hardening and precipitation hardening of the ω

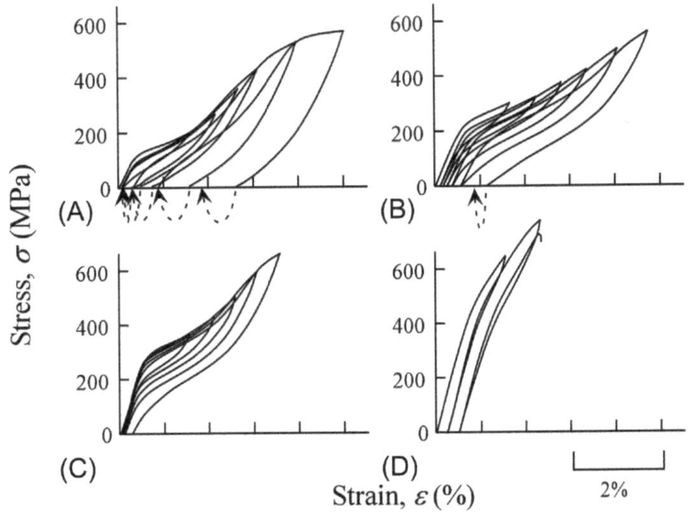

Fig. 4.11 Stress-strain curves obtained from strain increment cyclic tensile tests of (A) the Ti-26Nb alloy annealed at 873 K for 0.6 ks and the specimens aged at 573 K for (B) 1.8, (C) 3.6, and (D) 36 ks after annealing.

phase. It is also noted that the strengthening effect by aging is much pronounced in the annealed specimen in Fig. 4.11, when compared with the solution-treated specimen shown in Fig. 4.10. For example, the onset stress for the stress-induced martensitic transformation of the specimen aged for 3.6 ks after solution treatment at 1173 K is 160 MPa in Fig. 4.10, while that of the specimen aged for the same time after annealing at 873 K is 300 MPa in Fig. 4.11. Also, the specimen aged for 36 ks after annealing at 873 K exhibited a brittle fracture, while the specimen aged for 36 ks after solution treatment at 1173 K exhibited a good superelastic recovery. These results indicate that the low-temperature annealing after severe cold rolling enhanced the precipitation of the ω phase.

4.3.2 Effect of Nb Content on Age Hardening Response

Fig. 4.12 shows stress-strain curves obtained by cyclic loading-unloading tensile tests for Ti-27Nb and Ti-28Nb alloys annealed at 873 K for 0.6 ks and the specimens aged at 573 K for 3.6 ks after annealing [25]. Both alloys annealed at 873 K exhibited a good superelasticity up to the second cycle. It is seen that the onset stress for the stress-induced martensitic transformation of the Ti-27Nb alloy is lower than that of the Ti-28Nb alloy in the annealing condition. This is due to the Nb content dependence of the martensitic

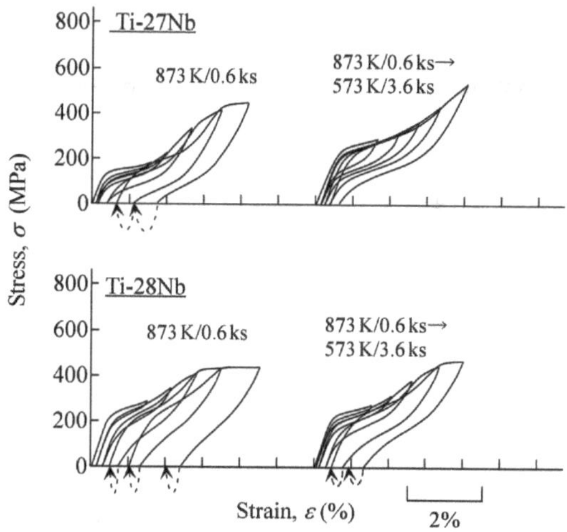

Fig. 4.12 Stress-strain curves obtained from strain increment cyclic tensile tests of Ti-27Nb and Ti-28Nb alloys annealed at 873 K for 0.6 ks and the specimens subsequently aged at 573 K for 3.6 ks.

transformation temperature. In both alloys, the aged specimens showed better superelastic properties. It should be noted that the increase in the onset stress by aging decreased with increasing Nb content, implying that the strengthening effect by the precipitation of the ω phase became weaker as the Nb content increased. This is due to the fact that the increase of the Nb content diminishes the precipitation of the ω phase because the stability of the β phase increases with increasing Nb content.

4.4 TEXTURE

Texture is one of the most important microstructural characteristics determining the shape memory properties. Particularly, the transformation strain is directly dependent on texture since the lattice distortion strain is strongly associated with the orientation of the crystals. The cold rolling textures and recrystallization textures reported in β Ti alloys [26–43] are listed in Table 4.1. The major texture components are dependent on not only the composition but also the rolling ratio.

4.4.1 Deformation Texture

Fig. 4.13 shows {011}, {002}, and {112} pole figures and an orientation distribution function (ODF) section of $\varphi_2=45°$ obtained from a

Table 4.1 Cold rolling and recrystallization textures of β Ti alloys

Alloy	Reduction ratio (%)	Deformation texture	Recrystallization texture	
Ti-10V-4.5Fe-1.5Al (wt%)	60	{001}⟨210⟩, {001}⟨130⟩, {015}⟨100⟩, {113}⟨332⟩, {011}⟨100⟩	{001} fiber, γ-fiber	[26]
Ti-24Nb-3Al	99	{001}⟨100⟩, {111}⟨112⟩	{112}⟨110⟩	[27]
Ti-22Nb-6Ta	90–95	{001}⟨110⟩	{112}⟨110⟩	[28]
Ti-35Nb-2Ta-3Zr (wt%)	20	{001}⟨110⟩, {111}⟨110⟩		[29]
	90	{001}⟨110⟩		
	99	{001}⟨110⟩	{112}⟨110⟩	[30]
Ti- (21–29) Nb-2Zr-0.7Ta-1.2O	98.5	{001} ⟨110⟩, {112}⟨110⟩		[31]
Ti-18Zr-16Nb	99	γ-Fiber		[32]
Ti-18Zr-18Nb	99	{001}⟨110⟩		
Ti-25Nb-10Ta-1Zr-0.2Fe (wt%)	30–80	γ-Fiber, {111}⟨110⟩, {111}⟨112⟩		[33]
Ti-35Nb-9Zr-6Mo-4Sn (wt%)	23–85	{001}⟨110⟩, {111}⟨110⟩		[34]
Ti-29Nb-9Ta-10Zr (wt%)	60	γ-Fiber, {001}, {001}⟨110⟩		[35]
Ti-25Ta-25Nb (wt%)	91.2	{001}⟨110⟩, γ-fiber, α-fiber		[36]
Ti-29Nb-9Ta-10Zr-0.2O (wt%)	90	{112}⟨111⟩, {111}⟨112⟩	{111} ⟨110⟩, γ-fiber	[37]

Table 4.1 Cold rolling and recrystallization textures of β Ti alloys—cont'd

Alloy	Reduction ratio (%)	Deformation texture	Recrystallization texture	
(Ti-Zr)-1.5Mo-3Sn	98.5	{113}⟨471⟩	{001}⟨110⟩	[38]
Ti-24Zr-10Nb-2Sn	98.5	{117}⟨110⟩	{001}⟨110⟩	[39]
Ti-18Zr-11Nb-3Sn	98.5		{001}⟨110⟩	[40]
Ti-26Nb-3Al	60–99	{001}⟨110⟩, {112}⟨110⟩	{112}⟨110⟩, {001}⟨110⟩	[41]
Ti-18Zr-4.5Nb-3Sn-2Mo	98.5	{113}⟨471⟩	{001}⟨110⟩	[42]
Ti-32.5Nb-6.8Zr-2.7Sn-0.3O (wt%)	25–90	{001}⟨110⟩, {112}⟨110⟩, {111}⟨110⟩		[43]

Ti-27Nb alloy cold-rolled with a reduction of 99% in thickness. The pole figures and corresponding ODF of the as-rolled specimen confirm the {001}⟨110⟩ texture. The {001}⟨110⟩ rolling texture has been reported to form in Ti-Nb-based alloys that have relatively large amounts of Nb such as Ti-24Nb-3Al [27], Ti-22Nb-6Ta [28], Ti-35Nb-2Ta-3Zr (wt%) [30], and Ti-23Nb-2Zr-0.7Ta-1.2O [31].

Cold rolling ratio is an important processing parameter to control the texture. It has been reported that as the cold rolling ratio increases the major component of deformation texture in β Ti alloys changes as follows: γ-fiber →{001}⟨110⟩ → α-fiber ({001}⟨110⟩ and {112}<110>) [33–35, 41]. Fig. 4.14 shows the effect of rolling ratio on the deformation texture in a Ti-26Nb-3Al alloy [41]. It is seen that the {001}⟨110⟩ is a dominant component and its intensity increases with increasing reduction ratio up to 97%. The {112}⟨110⟩ component appears in the specimens cold-rolled with a reduction >90%, and its intensity rapidly increases with increasing reduction ratio. It has been suggested that a pencil glide deformation mode is associated with the development of the {001}⟨110⟩ and {112}⟨110⟩ components [41, 44].

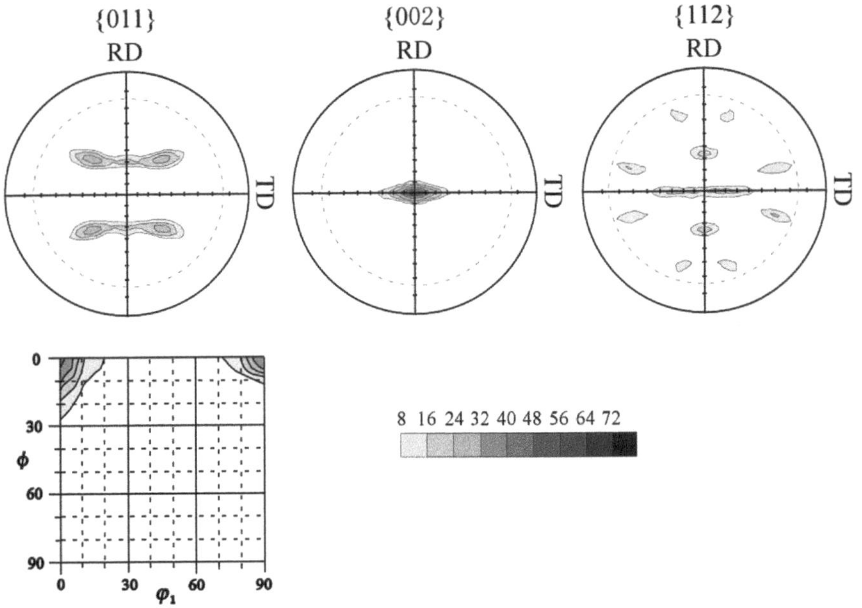

Fig. 4.13 Pole figures and an orientation distribution function (ODF) section of $\varphi_2 = 45°$ obtained from a Ti-27Nb alloy cold-rolled with a reduction of 99% in thickness.

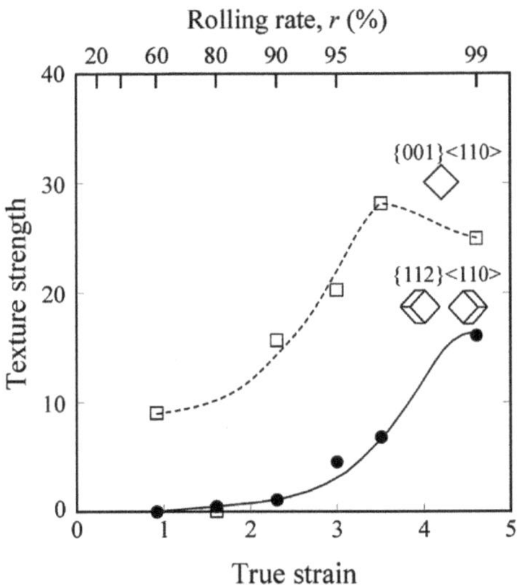

Fig. 4.14 Effect of rolling ratio on the intensity of major deformation texture components {001}<110> and {112}<110> in a Ti-26Nb-3Al alloy.

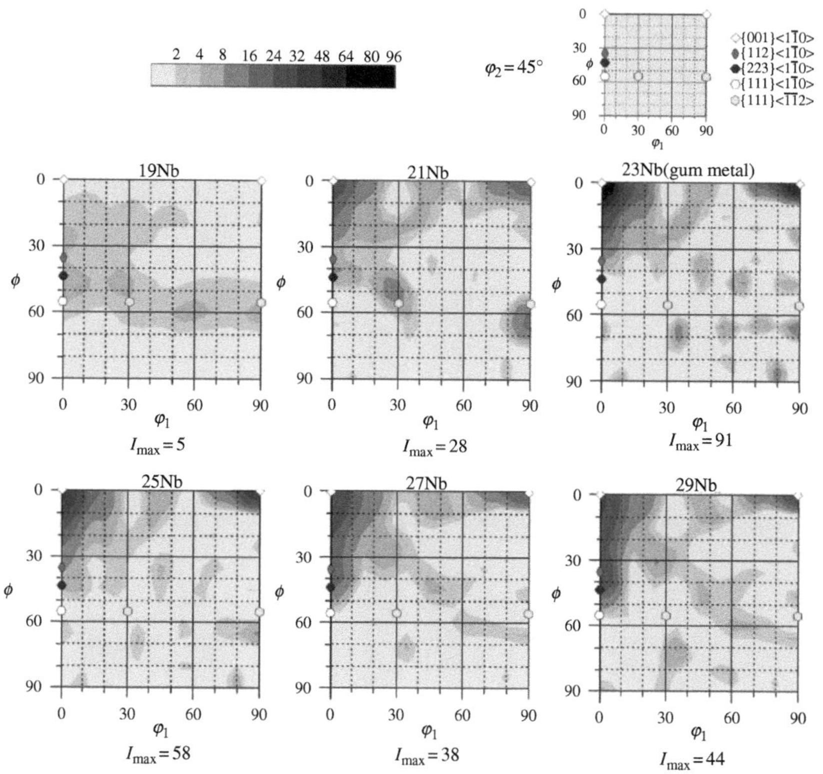

Fig. 4.15 ODF sections of $\varphi_2=45°$ for Ti-(19–29)Nb-2Zr-0.7Ta-1.2O alloys cold-rolled with a reduction of 98.5% in thickness.

The deformation texture is also sensitively dependent on the stability of the β phase. Fig. 4.15 shows the ODF sections of $\varphi_2=45°$ for Ti-(19–29)Nb-2Zr-0.7Ta-1.2O alloys cold-rolled with a reduction of 98.5% in thickness [31]. The Ti-19Nb-2Zr-0.7Ta-1.2O alloy, which has the lowest β stability, exhibits a weak γ-fiber texture. The Ti-(21–29)Nb-2Zr-0.7Ta-1.2O alloys reveal a partial α-fiber texture where the most dominant component is $\{001\}\langle110\rangle$. The intensity of each of major components $\{001\}\langle110\rangle$ and $\{112\}\langle110\rangle$ is plotted as a function of Nb content in Fig. 4.16 [31]. The intensity of the $\{001\}\langle110\rangle$ component increases drastically with increasing Nb content reaching maximum at 23 at.% and then decreases with further increasing Nb content. The $\{112\}\langle110\rangle$ component becomes stronger continuously with increasing Nb content. It has been reported that the Nb dependence of the deformation texture is due to the change of the deformation mode.

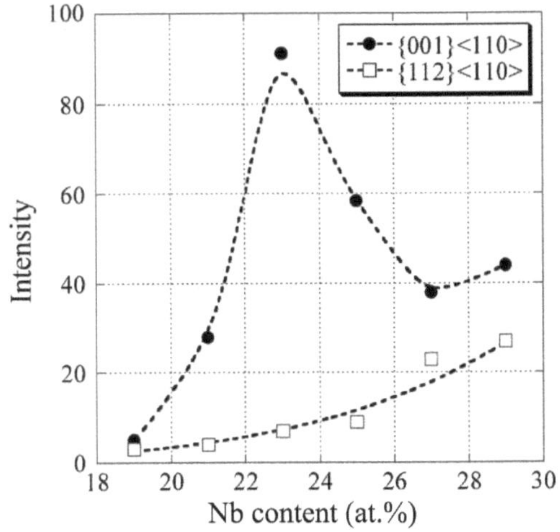

Fig. 4.16 Effect of Nb content on the intensity of major deformation texture components {001}<110> and {112}<110> in Ti-(19–29)Nb-2Zr-0.7Ta-1.2O alloys.

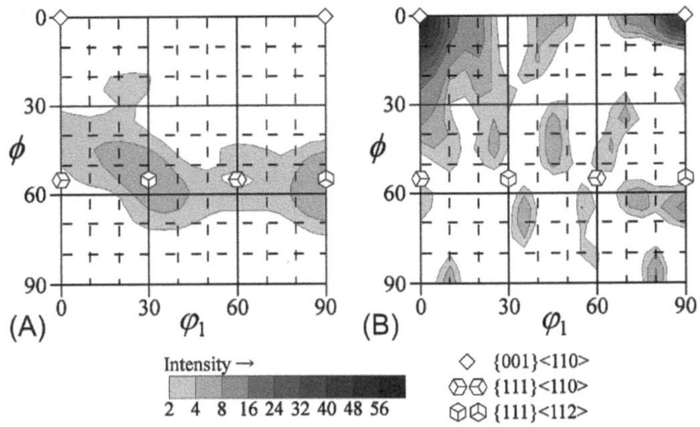

Fig. 4.17 ODF sections of $\varphi_2 = 45°$ for (A) Ti-18Zr-16Nb and (B) Ti-18Zr-18Nb alloys cold-rolled with a reduction of 99% in thickness.

The effect of Nb content on the texture evolution and microstructure has been also investigated in other β Ti alloys. Fig. 4.17 shows ODF sections of $\varphi_2 = 45°$ for Ti-18Zr-16Nb and Ti-18Zr-18Nb alloys cold-rolled with a reduction of 99% in thickness [32]. The Ti-18Zr-18Nb alloy exhibits a well-developed α-fiber texture with a maximum intensity at $\{001\}\langle 110\rangle$ that is

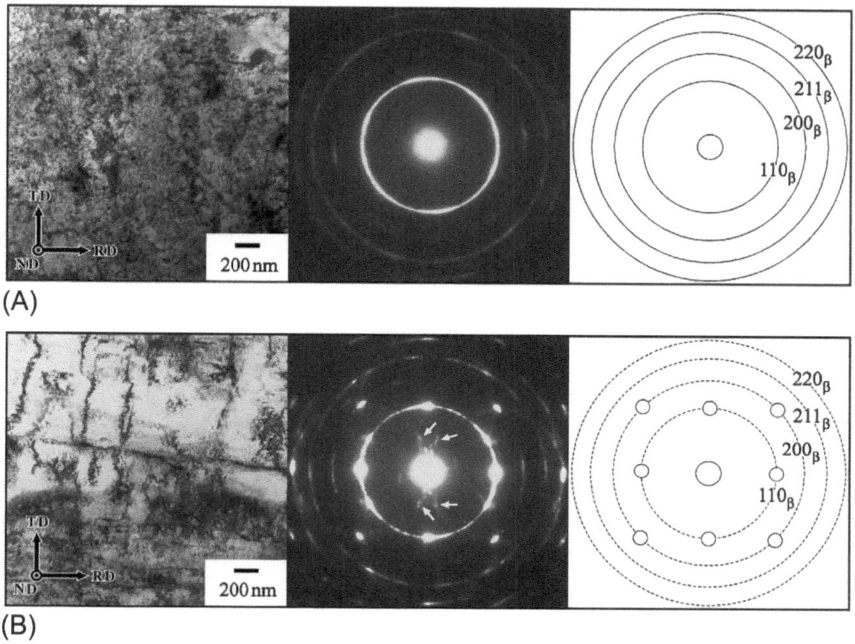

Fig. 4.18 TEM micrographs and the corresponding SADPs of (A) Ti-18Zr-16Nb and (B) Ti-18Zr-18Nb alloys cold-rolled with a reduction of 99% in thickness.

similar to those reported in other β Ti alloys such as Ti-26Nb and Ti-26Nb-3Al mentioned above. Only a small (2 at.%) decrease in the Nb content results in a completely different texture: the Ti-18Zr-16Nb alloy shows a weak γ-fiber texture with a maximum intensity at $\{111\}\langle 112\rangle$, in spite of a severe cold rolling of 99% reduction, similar to the Ti-19Nb-2Zr-0.7Ta-1.2O alloy. The deformation microstructures are also different in both alloys as shown in Fig. 4.18 [32]. The Ti-18Zr-16Nb alloy exhibits extremely fine structure with a continuous ring diffraction pattern. On the other hand, the Ti-18Zr-18Nb alloy reveals the dislocation cell structure with a high density of dislocations. The fine structure in the Ti-18Zr-16Nb alloy is suggested to be due to the stress-induced martensitic transformation during cold rolling and the complex deformation modes in the stress-induced α'' martensite phase. Similarly, a very weak deformation texture with a maximum intensity at $\{113\}\langle 471\rangle$ has been reported to form, although severe cold rolling of 98% reduction is applied, in Ti-Zr-based alloys such as Ti-18Zr-4.5Nb-3Sn-2Mo [42] and (Ti-Zr)-1.5Mo-3Sn [38].

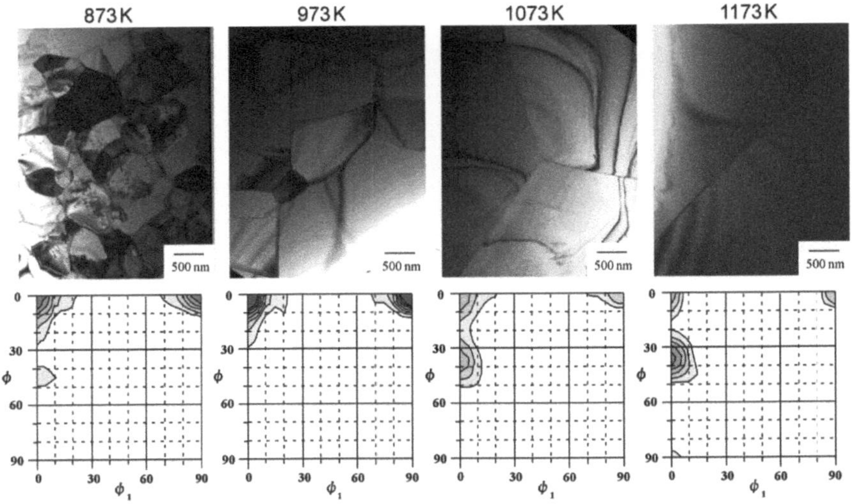

Fig. 4.19 TEM micrographs and ODF sections of $\varphi_2=45°$ for Ti-27Nb specimens annealed at different temperatures after cold rolling with a reduction of 99% in thickness.

4.4.2 Recrystallization Texture

Several recrystallization textures, such as $\{112\}\langle 110\rangle$, $\{001\}\langle 110\rangle$, and γ-fiber, have been reported in β Ti alloys. Particularly, the $\{112\}\langle 110\rangle$ recrystallization texture has been reported to form in the Ti-Nb alloys cold-rolled >90% in thickness, which have a strong $\{001\}\langle 110\rangle$ deformation texture. Fig. 4.19 shows TEM images and $\varphi_2=45°$ ODF sections for Ti-27Nb alloys annealed at different temperatures after cold rolling with a reduction of 99% in thickness. The annealing time was fixed at 1.8 ks. The specimens annealed at 873 and 973 K reveal a similar $\{001\}\langle 110\rangle$ texture observed in the as-rolled specimen (Fig. 4.13). As the annealing temperature increases, the $\{112\}<110>$ texture component becomes more pronounced. Fig. 4.20 shows the change in the intensity of each of the $\{001\}\langle 110\rangle$ and $\{112\}\langle 110\rangle$ components as a function of annealing temperature. The maximum orientation density of the specimens annealed at 873 and 973 K is higher than that of the as-rolled specimen, indicating that the deformation texture is intensified by recovery. A mixed texture composed of $\{001\}\langle 110\rangle$ and $\{112\}\langle 110\rangle$ with similar intensity is seen in the specimen annealed at 1073 K. The specimen annealed at 1173 K exhibits a strong $\{112\}\langle 110\rangle$ texture and a weak $\{001\}\langle 110\rangle$ texture, indicating that the recrystallization is almost completed. The recrystallization texture of

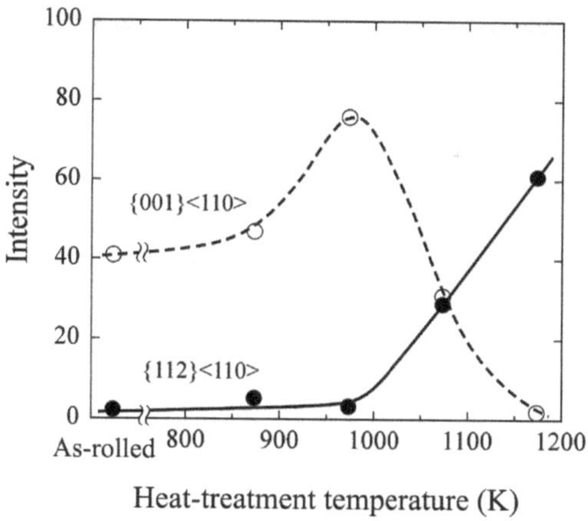

Fig. 4.20 Effect of annealing temperature on the intensity of major texture components {001}⟨110⟩ and {112}⟨110⟩ in the Ti-27Nb alloy.

{112}⟨110⟩ has been also reported to form in several alloys that show the {001}⟨110⟩ texture as the rolling texture [27, 28, 30, 41].

Fig. 4.21 shows the rolling ratio dependence of the recrystallization texture for the Ti-26Nb-3Al alloy subjected to solution treatment at 1273 K for 3.6 ks after the cold rolling at various thickness reductions [41]. The corresponding deformation textures are shown in Fig. 4.14. After the solution

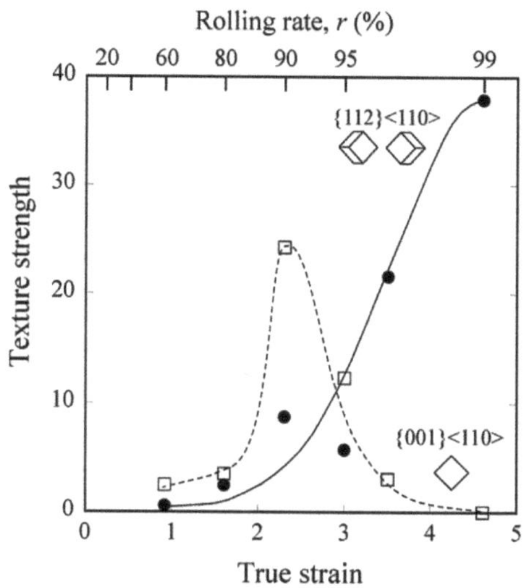

Fig. 4.21 Effect of rolling ratio on the intensity of major texture components {001}⟨110⟩ and {112}⟨110⟩ after the solution treatment for the Ti-26Nb-3Al alloy.

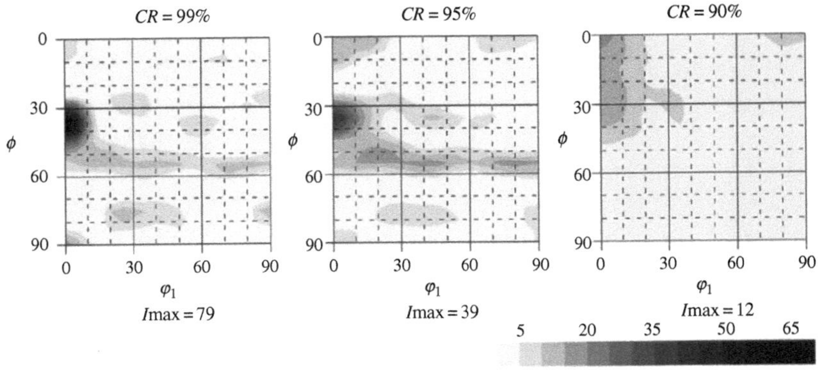

Fig. 4.22 ODF sections of $\varphi_2=45°$ of the specimens heat treated at 1173K for 1.8ks after the cold rolling with various rolling ratios of 90%, 95%, and 99%.

treatment, the intensity of {001}⟨110⟩ component is strongest at the rolling ratio of 90% and decreases with increasing rolling ratio. On the other hand, the intensity of the {112}⟨110⟩ component rapidly increased with increasing rolling ratio and becomes dominant as a recrystallization texture when the rolling ratio is higher than 95%. It is noted that only the {112}⟨110⟩ component is developed in the specimen cold-rolled up to 99%. The Ti-22Nb-6Ta alloy exhibits a similar trend in the cold rolling ratio dependence of the recrystallization texture as shown in Fig. 4.22 [28]. The specimens were heat treated at 1173K for 1.8ks after the cold rolling with various rolling ratios of 90%, 95%, and 99%. It is seen that a pronounced {112}⟨110⟩ recrystallization texture is developed in the 99% cold-rolled specimen. The 95% cold-rolled specimen exhibits a similar recrystallization texture although the intensity becomes weaker. A weak γ-fiber texture is also seen in the 95% and 99% cold-rolled specimens. On the other hand, the 90% cold-rolled specimen reveals the α-fiber texture extending from {001}⟨110⟩ to {112}⟨110⟩, indicating that the deformation texture is remained after the heat treatment at 1173K.

A strong {001}⟨110⟩ recrystallization texture has been reported in several β Ti alloys [38–40, 42] that have relatively low Nb content. Fig. 4.23 shows $\varphi_2=45°$ ODF sections of as-rolled (98.5% reduction in thickness) and annealed conditions of a Ti-18Zr-4.5Nb-3Sn-2Mo alloy [42]. As-rolled specimen exhibits a relatively weak texture with a maximum intensity at {113}⟨471⟩. Annealing treatment at 923K for 0.3ks intensifies the {113}⟨471⟩ component due to the recovery. In the specimens annealed

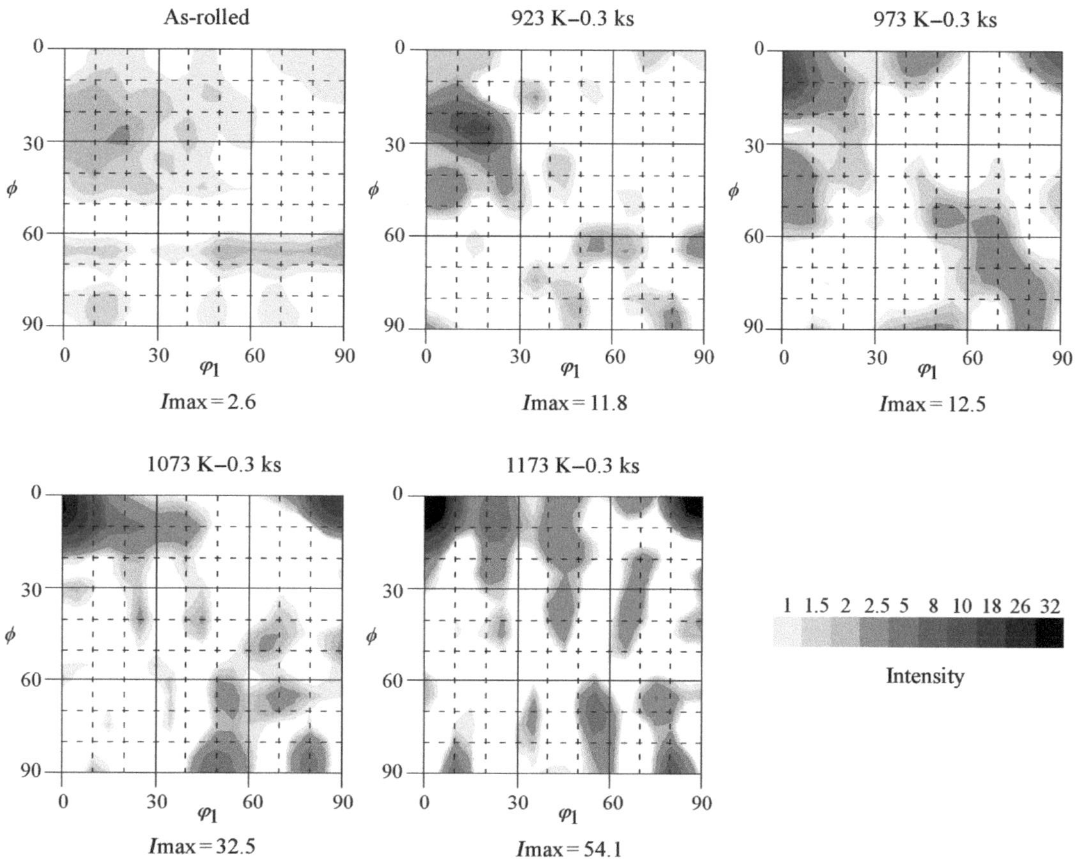

Fig. 4.23 ODF sections of $\varphi_2 = 45°$ of the as-rolled (98.5% reduction in thickness) and annealed conditions of a Ti-18Zr-4.5Nb-3Sn-2Mo alloy.

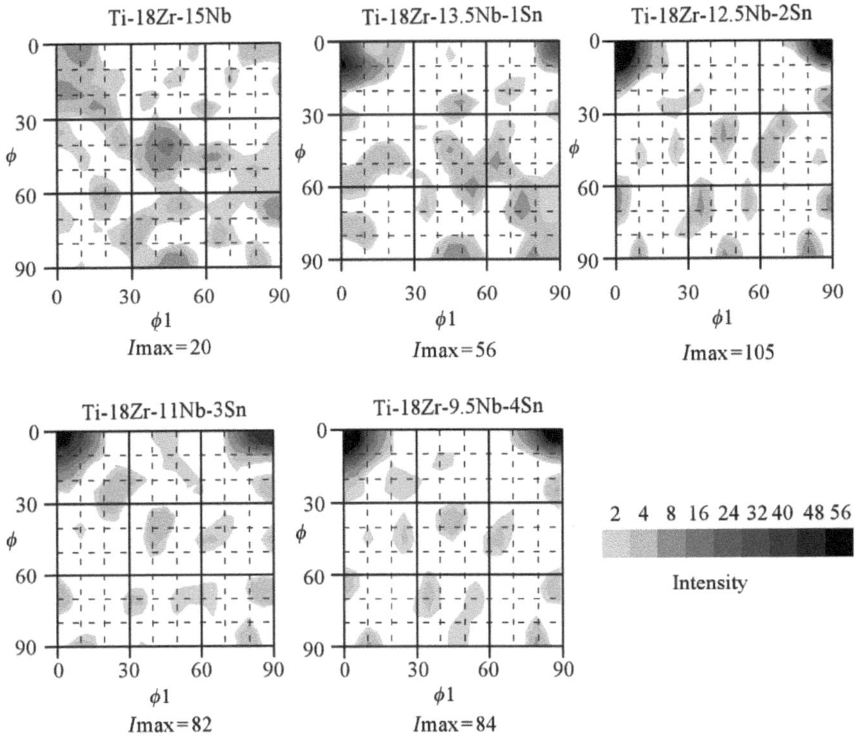

Fig. 4.24 ODF sections of Ti-18Zr-(9.5–15)Nb-(0–4)Sn alloy solution treated at 1173 K after cold rolling with a reduction of 98.5% in thickness.

at higher temperatures, the $\{001\}<110>$ texture is developed as a recrystallization texture. The peak intensity of the $\{001\}\langle 110\rangle$ component increases with increasing annealing temperature. A similar result has been reported in other Ti-Zr-based alloys such as Ti-24Zr-10Nb-2Sn [39], Ti-18Zr-11Nb-3Sn [40], and (Ti-Zr)-1.5Mo-3Sn [38].

4.4.3 Textures and Shape Memory Properties

Transformation strain is strongly dependent on the orientation of the crystals; therefore, shape memory properties are also affected by the texture of the alloys. The relation between texture and shape memory properties has been reported for several Ti alloys [27, 28, 38–40, 42, 45]. Fig. 4.24 shows the recrystallization textures of Ti-18Zr-(9.5–15)Nb-(0–4)Sn alloy solution treated at 1173 K after cold rolling with a reduction of 98.5% in thickness [40]. The recrystallization texture of Ti-18Zr-Nb alloys is changed by the addition of Sn. The Ti-18Zr-15Nb alloy displays a rather random texture with low intensity after recrystallization. On the other hand, a very strong

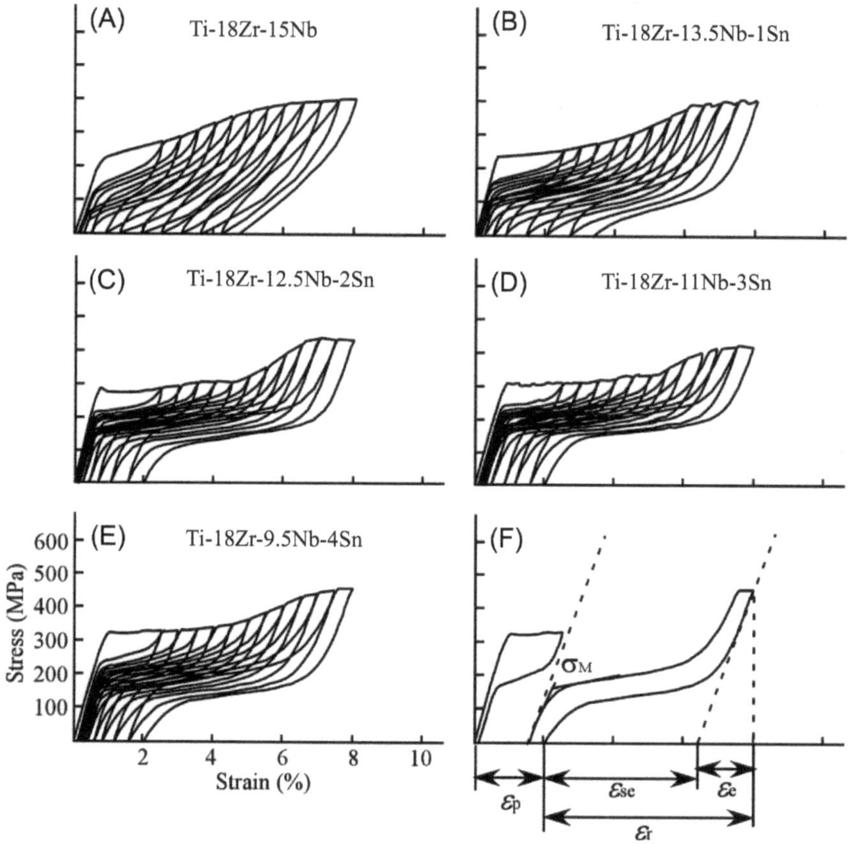

Fig. 4.25 Stress-strain curves obtained from strain increment cyclic tensile tests of Ti-18Zr-(9.5–15)Nb-(0–4)Sn alloys. (A) Ti-18Zr-15Nb, (B) Ti-18Zr-13.5Nb-1Sn, (C) Ti-18Zr-12.5Nb-2Sn, (D) Ti-18Zr-11Nb-3Sn, (E) Ti-18Zr-9.5Nb-4Sn, and (F) measurement scheme of superelastic properties.

{001}⟨110⟩ texture is developed as recrystallization texture in the Sn-added alloys. Fig. 4.25 shows the effect of Sn content on superelastic properties of Ti-18Zr-(9.5–15)Nb-(0–4)Sn alloys [40]. The cyclic tensile tests were carried out along the rolling direction (RD). It is noted that the Nb content is decreased with increasing Sn content in order to maintain the martensitic transformation temperature to exhibit superelasticity at room temperature. The Ti-18Zr-15Nb alloy exhibits superelasticity with a superelastic strain (ε_{se}) of 2.9%. As the Sn content increases, the superelastic strain increases up to about 5.0% and then decreases slightly.

Fig. 4.26 shows the inverse pole figures for the RD, i.e., the tensile direction of the specimens [40]. In the Ti-18Zr-15Nb alloy, the crystal directions

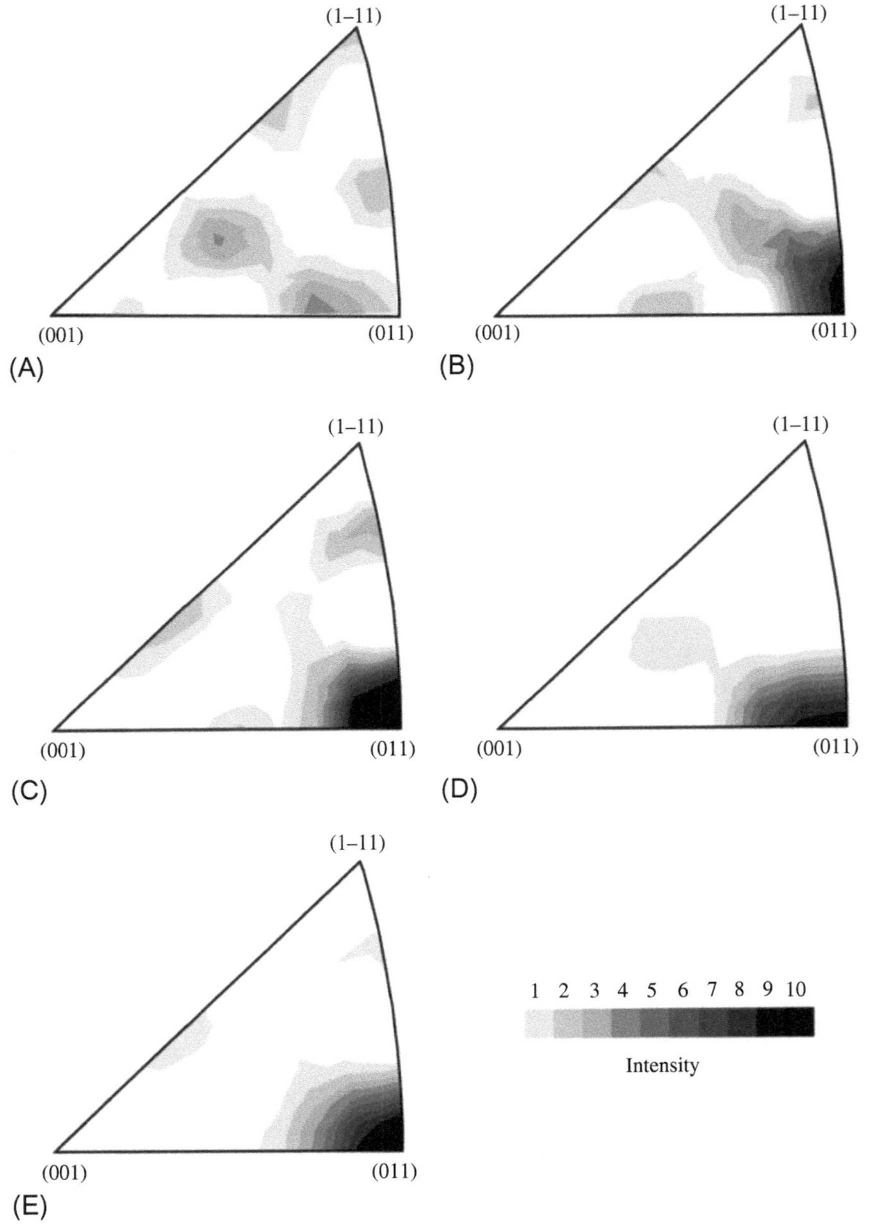

Fig. 4.26 Inverse pole figures for the RD of the Ti-18Zr-(9.5–15)Nb-(0–4)Sn alloys. (A) Ti-18Zr-15Nb, (B) Ti-18Zr-13.5Nb-1Sn, (C) Ti-18Zr-12.5Nb-2Sn, (D) Ti-18Zr-11Nb-3Sn, and (E) Ti-18Zr-9.5Nb-4Sn.

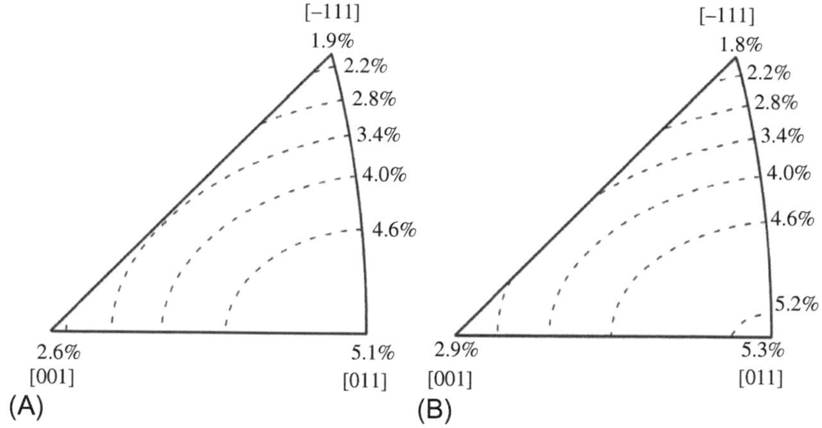

Fig. 4.27 Calculated transformation strains of the (A) Ti-18Zr-15Nb and (B) Ti-18Zr-11Nb-3Sn alloys plotted in a stand stereographic triangle.

seem to be distributed rather randomly with a relatively weak intensity. On the other hand, the inverse pole figures of Sn-added alloys show that the ⟨110⟩ directions are aligned parallel to the RD preferentially. The superelastic strain, i.e., transformation strain, is directly dependent on the crystal orientation. Fig. 4.27 compares the calculated transformation strains of the Ti-18Zr-15Nb and Ti-18Zr-11Nb-3Sn alloys plotted in a stand stereographic triangle [40]. In both alloys, a largest transformation strain is obtained along the [011] direction, and the transformation strain decreases continuously as the orientation deviates from the [011] direction. The transformation strains along the [011] direction are 5.1% and 5.4% for the Ti-18Zr-15Nb and Ti-18Zr-11Nb-3Sn alloys, respectively, indicating that the magnitude of transformation strain is not a major reason for the increase of the superelastic recovery strain in Sn-added alloys. From comparison of Figs. 4.26 and 4.27, it is clear that the primary reason of the increase in the superelastic strain by the increase of the Sn content is due to the desirable recrystallization texture: a very strong recrystallization texture of {001}⟨110⟩ produces the largest transformation strain along the rolling direction.

Fig. 4.28 shows the effect of annealing temperature on the superelastic properties of the Ti-18Zr-4.5Nb-3Sn-2Mo alloy [42]. It is seen that, in the specimen annealed at 923 K, the stress for the martensitic transformation raises as the applied strain increases, indicating that the progress of the martensitic transformation becomes difficult. The strain-hardening effect of the stress for the martensitic transformation becomes weaker as the annealing

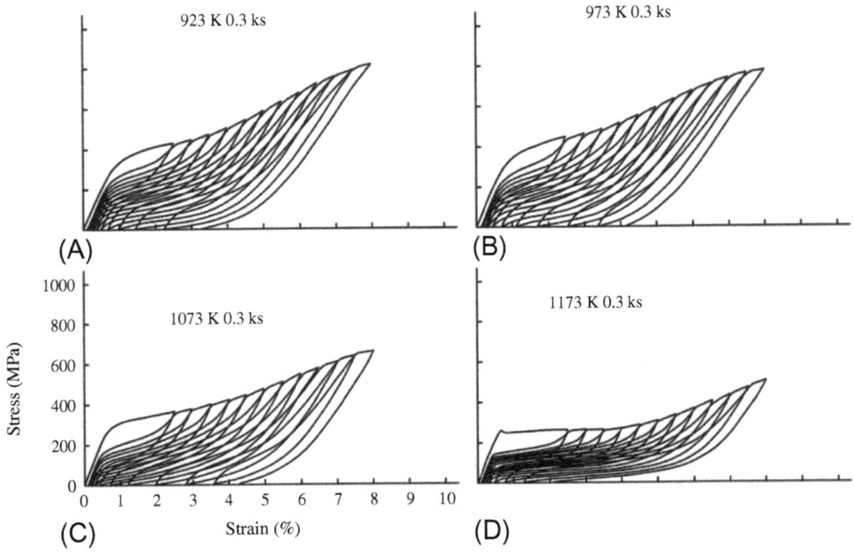

Fig. 4.28 Stress-strain curves obtained from strain increment cyclic tensile tests of Ti-18Zr-4.5Nb-3Sn-2Mo alloy specimens at (A) 923 K, (B) 973 K, (C) 1073 K, and (D) 1173 KK for 0.3 ks.

temperature increases. The specimen annealed at 1173 K exhibits a well-developed stress plateau region. It is also noticed that the onset stress for the stress-induced martensitic transformation decreases with increasing annealing temperature. Superelastic properties are improved remarkably as the annealing temperature increases. A large superelastic strain of about 5% is achieved in the specimen annealed at 1173 K. The well-developed stress plateau and the larger superelastic strain are attributed to the strong {001}<110> recrystallization texture. The inverse pole figures along the RD (Fig. 4.29 [42]) clearly show that the maximum intensity orientation is changed from $\langle 034 \rangle$ to $\langle 011 \rangle$ and the intensity of $\langle 011 \rangle$ direction increases with increasing annealing temperature.

A well-developed texture results in anisotropy of shape memory properties; the shape memory properties exhibit a strong loading direction dependence. Fig. 4.30 shows the tensile direction dependence of the shape memory properties for the Ti-27Nb specimens annealed at 873 K for 1.8 ks and at 1173 K for 1.8 ks, which represent strong $\{001\}\langle 110 \rangle$ and $\{112\}\langle 110 \rangle$ textures, respectively. For the specimen annealed at 873 K for 1.8 ks, the superelastic shape recovery was observed in all directions. It is noted that good superelasticity appears when the tensile direction is parallel to either

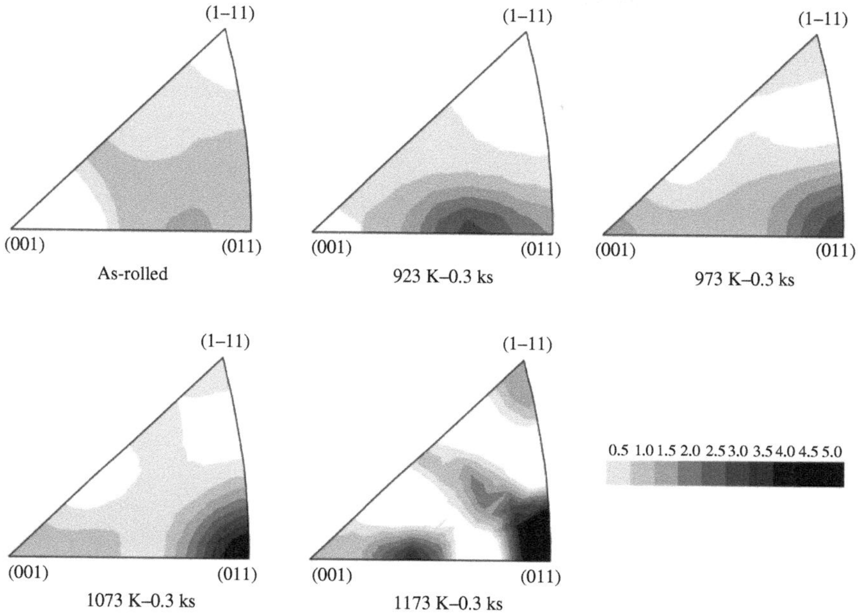

Fig. 4.29 Inverse pole figures for the rolling direction of the Ti-18Zr-4.5Nb-3Sn-2Mo specimens annealed at 923–1173 K for 0.3 ks.

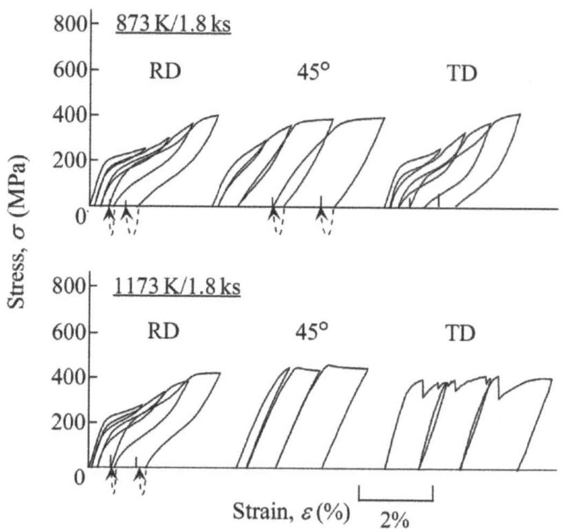

Fig. 4.30 Tensile direction dependence of the stress-strain curves for the Ti-27Nb specimens heat treated at 873 K for 1.8 ks and at 1173 K for 1.8 ks, respectively.

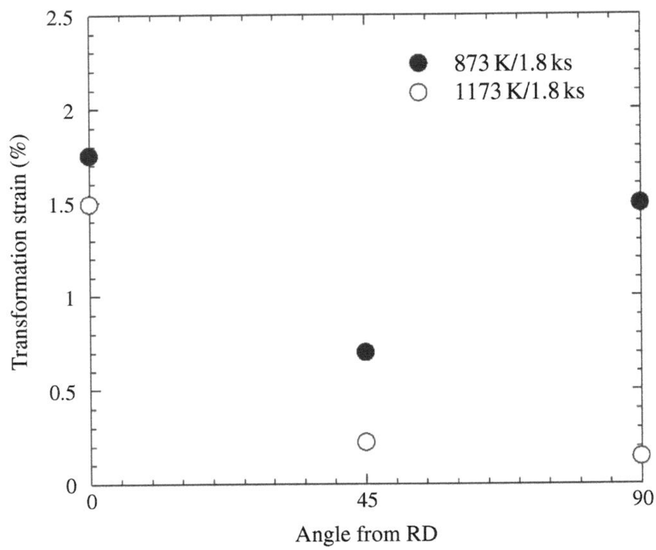

Fig. 4.31 The magnitude of the transformation strain plotted as a function of the angle from the RD for the Ti-27Nb specimens heat treated at 873 K for 1.8 ks and 1173 K for 1.8 ks, respectively.

RD or TD, and the stress-strain curves are very similar to each other. On the other hand, the specimen annealed at 1173 K for 1.8 ks exhibits good superelasticity only when the tensile direction is parallel to RD. Shape recovery is negligible when the tensile direction is 45° from RD and is parallel to TD. The apparent yield stress corresponds to the stress for inducing the martensitic transformation for the specimens annealed at 873 K, which exhibit superelasticity in all directions. For the specimen annealed at 1173 K, the apparent yield stress corresponds to the stress for inducing the martensitic transformation only when the specimen is loaded along RD. It is considered that the apparent yield stress of the specimen loaded along 45° from RD or TD corresponds to the critical stress for plastic deformation of the parent phase.

The magnitude of the transformation strain, i.e., maximum recovery strain due to the reverse transformation, is plotted as a function of the angle from RD, i.e., 0° (RD), 45°, and 90° (TD) from RD, in Fig. 4.31. For the specimen heat treated at 873 K, the largest transformation strain is obtained at RD. The transformation strain at TD exhibits slightly smaller value than that obtained at RD. The transformation strain exhibits minimum when the tensile direction is 45° from RD. For the specimen heat treated at 1173 K, the transformation strain obtained at RD is similar to that of the specimen heat

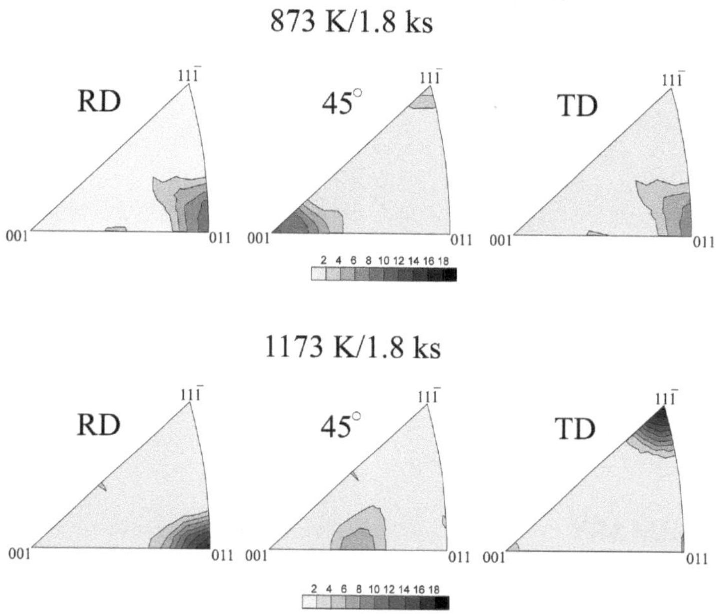

Fig. 4.32 Inverse pole figures of the Ti-27Nb specimens heat treated at 873 K for 1.8 ks and 1173 K for 1.8 ks, respectively, for the directions of RD, 45°, and TD.

treated at 873 K. On the other hand, when the tensile direction is TD or 45° from RD, very small transformation strain <0.2% was obtained for the specimen heat treated at 1173 K. Fig. 4.32 shows the inverse pole figures corresponding to three directions on the rolling plane. The inverse pole figures for RD and TD of the specimen heat treated at 873 K exhibit high axis density at [011], explaining that both specimens exhibit similar stress-strain curves. The inverse pole figure for 45° from RD shows that the [001] axis density presents a maximum peak. For the specimen heat treated at 1173 K, the inverse pole figures for RD and TD show high axis densities at [011] and [111], respectively. As shown in Fig. 4.33, the transformation strain of the Ti-27Nb alloy is largest along the [011] and decreases with deviation from the [011] orientation. As a result, it is concluded that the anisotropy of the shape memory properties can be successfully explained utilizing the texture information and the crystallography of martensitic transformation. It is also noted that both textures of $\{001\}\langle110\rangle$ and $\{112\}\langle110\rangle$ result in a large transformation strain along RD.

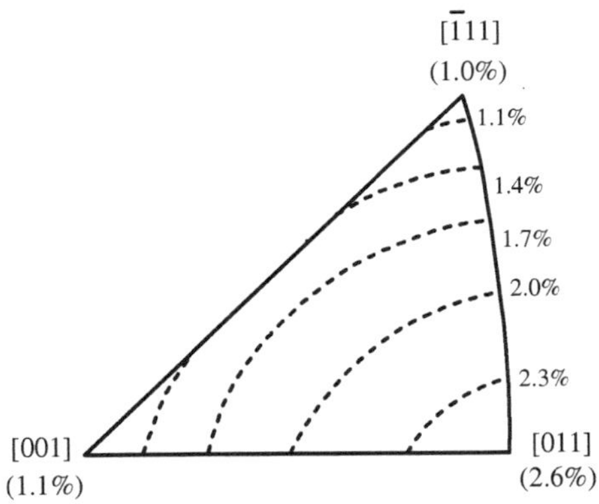

Fig. 4.33 Calculated transformation strains of the Ti-27Nb alloy plotted in a stand stereographic triangle.

4.5 SUMMARY

In order to bring out full potential of the β-type Ti-based shape memory alloys, it is essential to optimize the microstructure through thermomechanical treatment. For the β-type Ti-based shape memory alloys, strengthening and texture control are particularly important for improving the shape memory and superelastic properties. Annealing at relatively low temperatures below the recrystallization temperature after severe cold working is very effective for increasing the critical stress for plastic deformation of β Ti-based shape memory alloys. The optimum annealing temperature is dependent on the alloy composition. Generally, the critical stress for plastic deformation increases with decreasing annealing temperature due to the combination effect of the formation of fine α precipitates, high dislocation density, and small grain size. However, the alloys containing high levels of interstitial elements exhibit an opposite temperature dependence of the critical stress for plastic deformation due to the scavenging effect of α phase on the interstitial elements. Aging treatment is also very effective to increase the critical stress for plastic deformation. The formation of fine ω precipitates remarkably increases the critical stress for plastic deformation and improves superelastic properties; however, a large amount of ω precipitates hinders the martensitic transformation and decreases ductility, causing detrimental effects on the shape memory properties. Superelastic properties are strongly affected by texture. In β Ti-based shape memory alloys, $\{001\}\langle 110\rangle$, $\{112\}\langle 110\rangle$, and γ-fiber are the major recrystallization textures. Both $\{001\}\langle 110\rangle$ and $\{112\}\langle 110\rangle$ textures are favorable for the improvement of superelastic properties. By the combination of beneficial effects of large transformation strain and favorable grain alignment, some Ti-Zr-Nb-Sn alloys exhibit a large superelastic strain.

REFERENCES

[1] S. Miyazaki, K. Otsuka, Development of shape memory alloys, ISIJ Int. 29 (1999) 353–377.

[2] J.I. Kim, S. Miyazaki, Effect of nano-scaled precipitates on shape memory behavior of Ti–50.9at.%Ni alloy, Acta Mater. 53 (2005) 4545–4554.

[3] X. Wang, S. Kustov, K. Li, D. Schryvers, B. Verlinden, J. Van Humbeeck, Effect of nanoprecipitates on the transformation behavior and functional properties of a Ti–50.8 at.% Ni alloy with micron-sized grains, Acta Mater. 82 (2015) 224–233.

[4] S. Miyazaki, H.Y. Kim, H. Hosoda, Development and characterization of Ni-free Ti-base shape memory and superelastic alloys, Mater. Sci. Eng. A 438–440 (2006) 18–24.

[5] H.Y. Kim, S. Miyazaki, Martensitic transformation and superelastic properties of Ti-Nb base alloys, Mater. Trans. 56 (2015) 625–634.

[6] H.Y. Kim, S. Miyazaki, Several issues in the development of Ti–Nb-based shape memory alloys, Shap. Mem. Superelasticity 2 (2016) 380–390.

[7] J.I. Kim, H.Y. Kim, H. Hosoda, S. Miyazaki, Shape memory behavior of Ti–22Nb–(0.5–2.0)O (at%) biomedical alloys, Mater. Trans. 46 (2005) 852–857.

[8] M. Tahara, H.Y. Kim, H. Hosoda, S. Miyazaki, Shape memory effect and cyclic deformation behavior of Ti–Nb–N alloys, Funct. Mater. Lett. 2 (2009) 79–82.

[9] M. Tahara, H.Y. Kim, T. Inamura, H. Hosoda, S. Miyazaki, Effect of nitrogen addition on superelasticity of Ti–Zr–Nb alloys, Mater. Trans. 50 (2009) 2726–2730.

[10] M. Tahara, H.Y. Kim, H. Hosoda, T.H. Nam, S. Miyazaki, Effect of nitrogen addition and annealing temperature on superelastic properties of Ti–Nb–Zr–Ta alloys, Mater. Sci. Eng. A 527 (2010) 6844–6852.

[11] E.G. Obbard, Y.L. Hao, R.J. Talling, S.J. Li, Y.W. Zhang, D. Dye, R. Yang, The effect of oxygen on α'' martensite and superelasticity in Ti–24Nb–4Zr–8Sn, Acta Mater. 59 (2011) 112–125.

[12] A. Ramarolahy, P. Castany, F. Prima, P. Laheurtec, I. Péron, T. Gloriant, Microstructure and mechanical behavior of superelastic Ti–24Nb–0.5O and Ti–24Nb–0.5N biomedical alloys, J. Mech. Behav. Biomed. Mater. 9 (2012) 83–90.

[13] J.I. Kim, H.Y. Kim, T. Inamura, H. Hosoda, S. Miyazaki, Effect of annealing temperature on microstructure and shape memory characteristics of Ti–22Nb–6Zr(at%) biomedical alloy, Mater. Trans. 3 (2006) 505–512.

[14] S.J. Li, T.C. Cui, Y.L. Li, Y.L. Hao, R. Yang, Ultrafine-grained β-type titanium alloy with nonlinear elasticity and high ductility, Appl. Phys. Lett. 92 (2008)043128.

[15] W. Xu, X. Wu, M. Calin, M. Stoic, J. Eckert, K. Xia, Formation of an ultrafine-grained structure during equal-channel angular pressing of a β-titanium alloy with low phase stability, Scr. Mater. 60 (2009) 1012–1015.

[16] F. Sun, S. Nowak, T. Gloriant, P. Laheurte, A. Eberhardt, F. Prima, Influence of a short thermal treatment on the superelastic properties of a titanium-based alloy, Scr. Mater. 63 (2010) 1053–1056.

[17] F. Sun, Y.L. Hao, S. Nowak, T. Gloriant, P. Laheurte, F. Prima, A thermomechanical treatment to improve the superelastic performances of biomedical Ti–26Nb and Ti–20Nb–6Zr (at%) alloys, J. Mech. Behav. Biomed. Mater. 4 (2011) 1864–1872.

[18] L.W. Ma, H.S. Cheng, C.Y. Chung, Effect of thermo-mechanical treatment on superelastic behavior of Ti–19Nb–14Zr (at.%) shape memory alloy, Intermetallics 32 (2013) 44–50.
[19] S. Dubinskiy, V. Brailovski, S. Prokoshkin, V. Pushin, K. Inaekyan, V. Sheremetyev, M. Petrzhik, M. Filonov, Structure and properties of Ti–19.7Nb–5.8Ta shape memory alloy subjected to thermomechanical processing including aging, J. Mater. Eng. Perform. 22 (2013) 2656–2664.
[20] D.C. Zhang, C.G. Tan, D.M. Tang, Y. Zhang, J.G. Lin, C.E. Wen, Effect of thermo-mechanical treatment on the superelasticity of Ti–7.5Nb–4Mo–2Sn biomedical alloy, Mater. Sci. Eng. C 44 (2014) 76–86.
[21] E. Takahashi, T. Sakurai, S. Watanabe, N. Masahashi, S. Hanada, Effect of heat treatment and Sn content on superelasticity in biocompatible TiNbSn alloys, Mater. Trans. 43 (2002) 2978–2983.
[22] T. Maeshima, S. Ushimaru, K. Yamauchi, M. Nishida, Effects of Sn content and aging conditions on superelasticity in biomedical Ti–Mo–Sn alloys, Mater. Trans. 47 (2006) 513–517.
[23] H.Y. Kim, Y. Ikehara, J.I. Kim, H. Hosoda, S. Miyazaki, Martensitic transformation, shape memory effect and superelasticity of Ti–Nb binary alloys, Acta Mater. 54 (2006) 2419–2429.
[24] H.Y. Kim, Y. Al-Zain, T. Inamura, H. Hosoda, S. Miyazaki, in: Effect of omega phase on shape memory properties of Ti-base alloys, Proc. 12th World Conference on Titanium (Ti-2011), 2012, , pp. 1110–1113.
[25] H.Y. Kim, J.I. Kim, T. Inamura, H. Hosoda, S. Miyazaki, Effect of thermo-mechanical treatment on mechanical properties and shape memory behavior of Ti–(26–28) at.% Nb alloys, Mater. Sci. Eng. A 438–440 (2006) 839–843.
[26] A.K. Singh, A. Bhattacharjee, A.K. Gogia, Microstructure and texture of rolled and annealed β titanium alloy Ti–10V–4.5Fe–1.5Al, Mater. Sci. Eng. A 270 (1999) 225–230.
[27] T. Inamura, Y. Fukui, H. Hosoda, K. Wakashima, S. Miyazaki, Relationship between texture and macroscopic transformation strain in severely cold-rolled Ti-Nb-Al superelastic alloy, Mater. Trans. 45 (2004) 1083–1089.
[28] H.Y. Kim, T. Sasaki, K. Okutsu, J.I. Kim, T. Inamura, H. Hosoda, S. Miyazaki, Texture and shape memory behavior of Ti–22Nb–6Ta alloy, Acta Mater. 54 (2006) 423–433.
[29] L.Q. Wang, W.J. Lu, J.N. Qin, F. Zhang, D. Zhang, Texture and superelastic behavior of cold-rolled TiNbTaZr alloy, Mater. Sci. Eng. A 491 (2008) 372–377.
[30] L.Q. Wang, W.J. Lu, J.N. Qin, F. Zhang, D. Zhang, The characterization of shape memory effect for low elastic modulus biomedical β-type titanium alloy, Mater. Charact. 61 (2010) 535–541.
[31] Y. Nii, H.Y. Kim, S. Miyazaki, Effect of Nb content on plastic deformation behavior and deformation textures of Ti-Nb-Zr-ta-O alloy, J. Jpn. Inst. Metals 72 (2008) 970–974.
[32] H. Tobe, H.Y. Kim, S. Miyazaki, Effect of Nb content on deformation textures and mechanical properties of Ti-18Zr-Nb biomedical alloys, Mater. Trans. 50 (2009) 2721–2725.
[33] Y.F. Xu, D.Q. Yi, H.Q. Liu, X.Y. Wu, B. Wang, F.L. Yang, Effects of cold deformation on microstructure, texture evolution and mechanical properties of Ti–Nb–Ta–Zr–Fe alloy for biomedical applications, Mater. Sci. Eng. A 547 (2012) 64–71.
[34] S. Dai, Y. Wang, F. Chen, X. Yu, Y. Zhang, Effects of cold deformation on microstructure and mechanical properties of Ti–35Nb–9Zr–6Mo–4Sn alloy for biomedical applications, Mater. Sci. Eng. A 575 (2013) 35–40.

[35] V.D. Cojocaru, D. Raducanu, T. Gloriant, D.M. Gordin, I. Cinca, Effects of cold-rolling deformation on texture evolution and mechanical properties of Ti–29Nb–9Ta–10Zr alloy, Mater. Sci. Eng. A 586 (2013) 1–10.

[36] V.D. Cojocaru, D. Raducanu, D.M. Gordin, I. Cinca, Texture in ultra-strength Ti–25Ta–25Nb alloy strips, J. Alloys Compd. 576 (2013) 170–176.

[37] V.D. Cojocaru, D. Raducanu, D.M. Gordin, I. Cinca, I. Thibon, A. Caprarescu, Recrystallization temperature influence upon texture evolution of a SPD processed Ti-Nb-Ta-Zr-O alloy, IOP Conf. Ser. Mater. Sci. Eng. 63 (2014) 012151.

[38] M.F. Ijaz, H.Y. Kim, H. Hosoda, S. Miyazaki, Superelastic properties of biomedical (Ti–Zr)–Mo–Sn alloys, Mater. Sci. Eng. C 48 (2014) 11–20.

[39] L.L. Pavón, H.Y. Kim, H. Hosoda, S. Miyazaki, Effect of Nb content and heat treatment temperature on superelastic properties of Ti–24Zr–(8–12)Nb–2Sn alloys, Scr. Mater. 95 (2015) 46–49.

[40] J. Fu, A. Yamamoto, H.Y. Kim, H. Hosoda, S. Miyazaki, Novel Ti-base superelastic alloys with large recovery strain and excellent biocompatibility, Acta Biomater. 17 (2015) 56–67.

[41] T. Inamura, R. Shimizu, H.Y. Kim, S. Miyazaki, H. Hosoda, Optimum rolling ratio for obtaining {001}<110> recrystallization texture in Ti–Nb–Al biomedical shape memory alloy, Mater. Sci. Eng. C 61 (2016) 499–505.

[42] J. Fu, H.Y. Kim, S. Miyazaki, Effect of annealing temperature on microstructure and superelastic properties of a Ti-18Zr-4.5Nb-3Sn-2Mo alloy, J. Mech. Behav. Biomed. Mater. 65 (2017) 716–723.

[43] C. Lan, Y. Wu, L. Guo, F. Chen, Effects of cold rolling on microstructure, texture evolution and mechanical properties of Ti-32.5Nb-6.8Zr-2.7Sn-0.3O alloy for biomedical applications, Mater. Sci. Eng. A 690 (2017) 170–176.

[44] I.L. Dillamore, H. Katoh, A comparison of the observed and predicted deformation textures in cubic metals, Met. Sci. 8 (1974) 21–27.

[45] Y. Fukui, T. Inamura, H. Hosoda, K. Wakashima, S. Miyazaki, Mechanical properties of a Ti–Nb–Al shape memory alloy, Mater. Trans. 45 (2004) 1077–1082.

CHAPTER 5

Unique Properties of Metastable Beta Ti Alloys Related to Martensitic Transformation

Contents

5.1 INTRODUCTION

Over the last decades, β-type Ti alloys have attracted a great deal of attention because of their superior properties such as high strength, good formability, high corrosion resistance, and excellent biocompatibility. Particularly metastable β-type Ti alloys have been paid increasing attention due to their promising applications in aerospace and biomedical industries, and many new alloys have been developed. "Gum metal" is one of the unique β-type Ti alloys, characterized by low Young's modulus, high strength, high elastic deformability, and high plastic deformability at room temperature [1–4]. Gum metal also exhibits multiple anomalous properties such as nonlinear elasticity, Invar-like behavior, Elinvar-like behavior, and very low work-hardening rate [1–4]. There have been extensive efforts to elucidate the origin of such unique properties of gum metal [5–17]. Although the

Ni-free Ti-based Shape Memory Alloys
https://doi.org/10.1016/B978-0-12-809401-3.00005-7

mechanisms of these unique properties are not fully understood yet, it has become clear that the unique characteristics are closely related to the martensitic transformation [18–25].

$\{332\}\langle 113\rangle$ twinning has also gained attention as a unique deformation mechanism in the metastable β Ti alloys because it is effective in enhancing the work-hardening ability due to twinning-induced plasticity (TWIP) effect [26–28]. There have been many attempts to clarify the reason why the $\{332\}\langle 113\rangle$ twinning occurs preferentially in the metastable β Ti alloys instead of the $\{112\}<111>$ twinning that is the dominating twinning mode in most of metals and alloys with a bcc structure [29–34]. Recent researches show that the martensitic transformation plays a key role in the formation of the $\{332\}\langle 113\rangle$ twinning [35–38].

Low Young's modulus of metastable β Ti alloys is particularly attractive for biomedical applications since it is helpful for reducing stress shielding effect caused by Young's modulus mismatch between metallic implants and bones [39–42]. Over the last decade, there have been significant efforts to develop β Ti alloys that have Young's modulus much closer to the cortical bone than the traditional metallic implant alloys. Low Young's moduli <70 GPa have been successfully achieved in several alloys through the alloy design considering the β-phase stability [39–42].

In this chapter, some unique properties of gum metal, such as nonlinear elasticity and Invar-like thermal expansion behavior, are reviewed, and the underlying mechanisms and the role of the martensitic transformation are discussed. In addition, the origin of the $\{332\}$ twinning that is a characteristic deformation twinning of metastable β Ti alloys is verified. Also, the similarities and differences between superelasticity and low Young's modulus are discussed.

5.2 UNIQUE PROPERTIES OF GUM METAL

Gum metals have attracted great attention from both academic and industrial fields due to their multifunctional properties such as excellent cold workability, low elastic modulus, high strength, Invar-like behavior, and Elinvar-like behavior [1–4]. Many discussions have been made on the mechanisms of the unique properties of gum metals, and it has become clear that those peculiar and unique properties are closely related to phase stability and martensitic transformation [18–25].

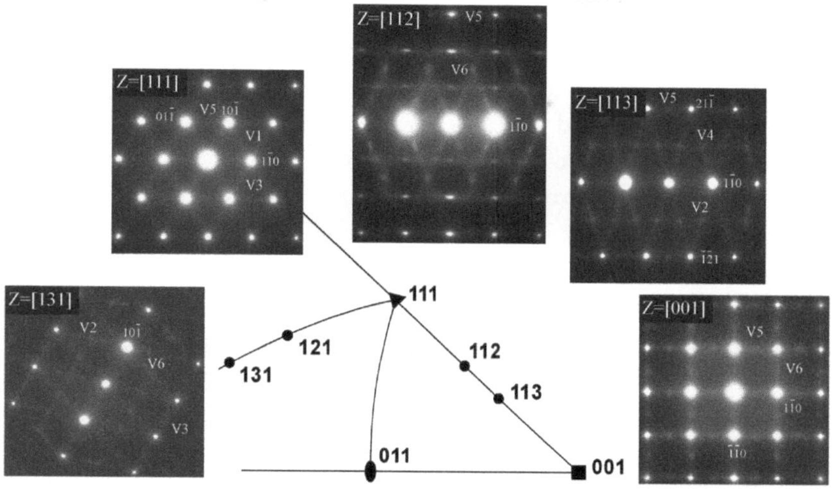

Fig. 5.1 A series of selected area diffraction patterns of a Ti-23Nb-2Zr-0.7Ta-1.2O (at.%) alloy taken along various zone axes.

5.2.1 Lattice Modulation and Nanodomain Structure

Fig. 5.1 shows a series of selected area diffraction patterns along various zone axes obtained from a Ti-23Nb-2Zr-0.7Ta-1.2O (at.%) alloy that has been extensively investigated as a typical example of gum metals [23]. Besides the fundamental spots from the β phase and the spots from the ω phase, diffuse spots and streaks are seen in all selected diffraction patterns. These diffuse streaks have been confirmed to be caused by the $\{110\}\langle 1\bar{1}0\rangle$ type lattice modulation associated with the shuffling mode of the martensitic transformation from β to α″. As mentioned in Chapter 3, the origin of the $\{110\}\langle 1\bar{1}0\rangle$ lattice modulation is closely related to interstitial oxygen atoms. The oxygen atoms, which occupy octahedral sites in the bcc structure, generate the stress field along $\langle 001\rangle$ direction. Accordingly, the $\{110\}\langle 1\bar{1}0\rangle$ lattice modulation is considered to be due to the relaxation of the local stress field. Niu and Geng [43] demonstrated, by a first-principle density functional theory (DFT) study, that O atoms impose a suppression effect on the martensitic transformation from β to α″, and the lattice distortion induced by O atoms generates a semi-α″ structure. A more detailed explanation on the origin of lattice modulation is given in Chapter 3. The $\{110\}\langle 1\bar{1}0\rangle$ lattice modulation has six equivalent variants, and the existence of diffuse streaks that originated from all six variants of $\{110\}\langle 1\bar{1}0\rangle$ lattice modulation is validated in Fig. 5.1. It is also seen in Fig. 5.1 that the diffuse streaks corresponding to the different variants of the lattice modulation have similar intensity, indicating that six variants of the $\{110\}\langle 1\bar{1}0\rangle$ lattice modulation exist with similar probabilities. Fig. 5.2 shows dark-field TEM micrographs obtained from intensity maxima of streaks corresponding to shuffling modes V5 and V6 [23]. Nanometer-sized domain structure is clearly seen in both dark-field TEM micrographs. The increase of oxygen content has been confirmed to increase the intensity of streaks and the amount of nanodomains in Ti-23Nb-2Zr-0.7Ta-O alloys [24].

5.2.2 Nonlinear Elastic Behavior

Fig. 5.3 shows the effect of oxygen content on deformation behavior of Ti-23Nb-2Zr-0.7Ta-(0.3–1.8)O (at.%) alloys annealed at 1023 K for 0.6 ks after cold rolled up to a 98.5% reduction in thickness [25]. The stress-strain curves were obtained upon tensile loading and unloading tests at RT. The Ti-23Nb-2Zr-0.7Ta-0.3O alloy exhibits superelasticity with a stress hysteresis about 150 MPa. The XRD profiles (Fig. 5.3B) clearly show that the superelastic recovery in the Ti-23Nb-2Zr-0.7Ta-0.3O alloy is due to the stress-induced martensitic transformation and its reverse transformation during loading and unloading, respectively. This result implies that gum metals exhibit conventional superelastic behavior when the oxygen content is low. The onset stress for the stress-induced martensitic transformation increases with increasing oxygen content. As the oxygen content increases, the stress hysteresis decreases, and the deformation mode changes to nonlinear elastic behavior. The in situ XRD profiles of Ti-23Nb-2Zr-0.7Ta-1.5O alloy show no distinct peak from the martensite phase. This is

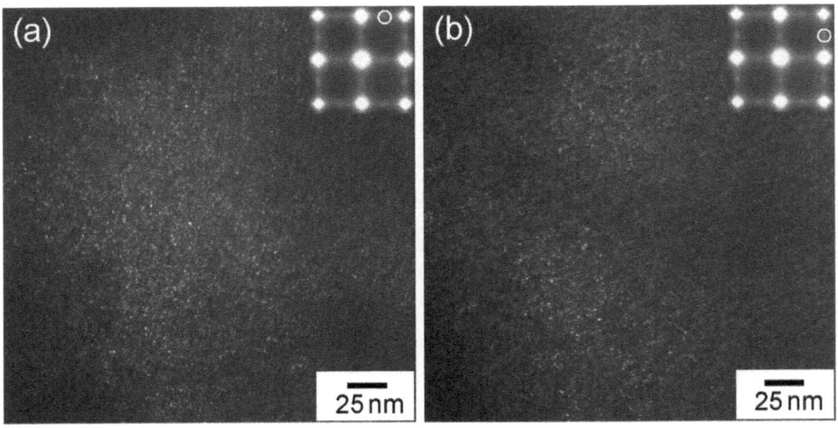

Fig. 5.2 Dark-field TEM micrographs obtained from intensity maxima of streaks corresponding to shuffling modes (A) V5 and (B) V6.

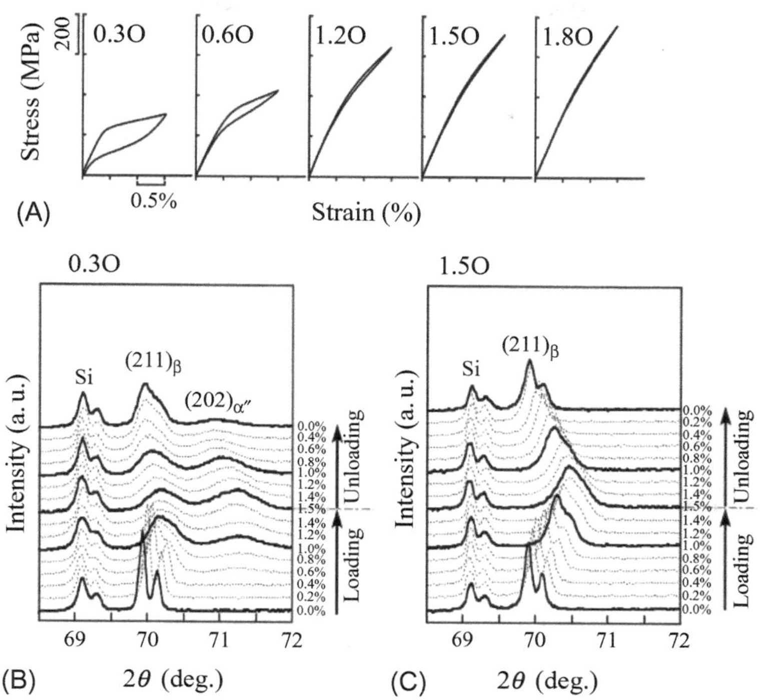

Fig. 5.3 (A) Stress-strain curves upon tensile loading and unloading for Ti-23Nb-2Zr-0.7Ta-(0.3–1.8)O (at.%) alloys and in situ XRD profiles of (B) 0.3O and (C) 1.5O alloys.

essentially the same behavior seen in the Ti-Nb-O alloys, and the same mechanism is proposed to describe the effect of oxygen content: the oxygen facilitates the formation of nanodomain structure, and it hinders a long-range martensitic transformation.

Fig. 5.4 shows the stress-strain curves of the Ti-23Nb-2Zr-0.7Ta-(0.3–1.8)O (at.%) alloys obtained upon tensile loading and unloading at various test temperatures [25]. For the 0.3–0.6O alloys, the apparent yield stress, indicated by an arrow in each stress-strain curve on loading, decreases, and the stress plateau becomes flatter with decreasing temperature from 313 to 233 K. This temperature dependence of the apparent yield stress is explained by the Clausius-Clapeyron relationship of the stress-induced martensitic transformation. The increase in the apparent yield stress with further decrease of temperature from 233 to 193 K is due to the formation of athermal ω phase that suppresses the martensitic transformation. For the alloys with higher oxygen contents, although the stress plateau region does not appear clearly, the change of slope at the yield point becomes more

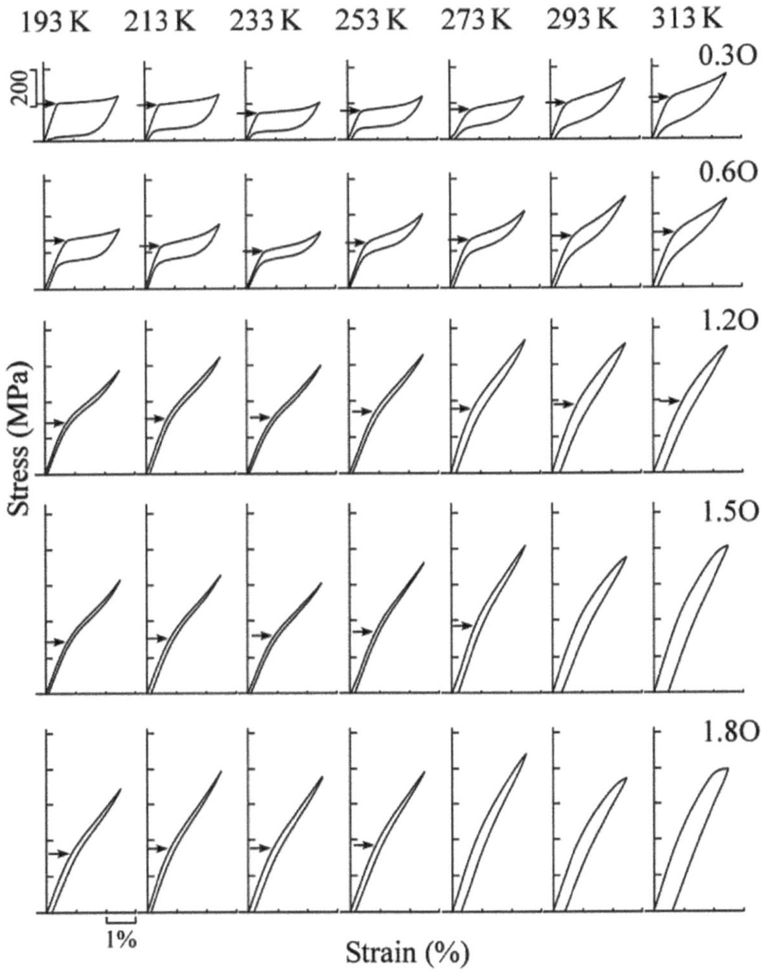

Fig. 5.4 Stress-strain curves of the Ti-23Nb-2Zr-0.7Ta-(0.3–1.8)O (at.%) alloys obtained at various test temperatures.

pronounced at lower temperatures. Furthermore, the apparent yield stress decreases almost linearly with decreasing temperature in the temperature range investigated, satisfying the Clausius-Clapeyron relationship. These results also imply that the deformation behavior of gum metals is associated with the martensitic transformation. A stress-induced reversible phase transformation in the deformation of gum metal was also confirmed by Talling et al. [20, 44] using in situ synchrotron x-ray diffraction (SXRD).

5.2.3 Invar-Like Behavior

Fig. 5.5 shows thermal expansion curves of a gum metal with a composition of Ti-23Nb-2Zr-0.7Ta-1.2O (at.%) cold rolled to a reduction in thickness of 98.5% and annealed at 1023 K for 0.6 ks [23]. The annealed specimens exhibited ordinary positive thermal expansion curves in both rolling direction (RD) and transverse direction (TD) with a thermal expansion coefficient about $7 \times 10^{-6}\ K^{-1}$ that is similar to thermal expansion coefficients of other β Ti alloys. In the as-rolled specimens, thermal expansion curves are greatly varied depending on the orientation. A very low thermal expansion close to zero is observed in the as-rolled specimen measured along the RD. On the other hand, a positive thermal expansion curve similar to those of the annealed specimens is observed along the TD. These results indicate that the Invar-like behavior of gum metals is achieved only along the RD in the cold-rolled specimen.

In order to explain the Invar-like behavior in the as-rolled specimen, the "preferential growth of nanodomain" model has been proposed. Fig. 5.6 shows a typical SADP with the [001] zone axis obtained from the normal plane of the as-rolled specimen [23]. It is reasonable that {001} planes are most frequently observed along the normal direction of the cold-rolled

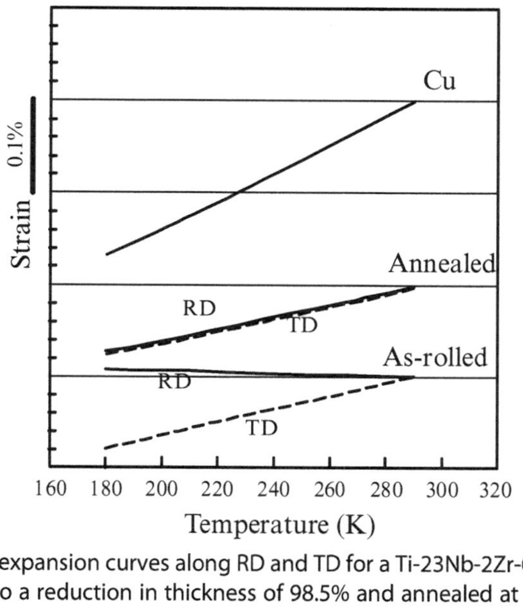

Fig. 5.5 Thermal expansion curves along RD and TD for a Ti-23Nb-2Zr-0.7Ta-1.2O (at.%) alloy cold rolled to a reduction in thickness of 98.5% and annealed at 1023 K for 0.6 ks. The thermal expansion curve of Cu is also plotted for reference.

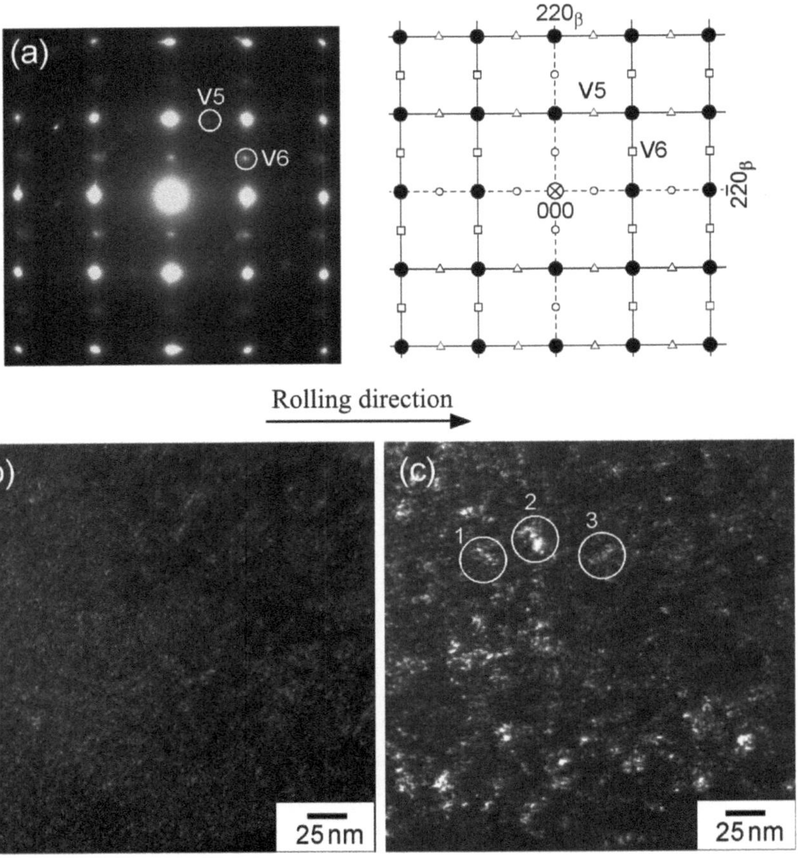

Fig. 5.6 (A) A selected diffraction pattern of a Ti-23Nb-2Zr-0.7Ta-1.2O (at.%) alloy cold rolled to a reduction in thickness of 98.5% and dark-field TEM micrographs obtained from intensity maxima of streaks corresponding to shuffling modes (B) V5 and (C) V6.

specimen because of a strong cold rolling texture of $\{001\}\langle 110\rangle$. Unlike the annealed specimen (Fig. 5.2), the intensities of diffuse spots on the $[\bar{1}10]^*$ and $[110]^*$, corresponding to the nanodomain variants V5 and V6, respectively, are different in the as-rolled specimen. The diffuse spots on the $[110]^*$ are more noticeable, while the diffuse spots on the $[\bar{1}10]^*$ are fainter. The preferential growth of the nanodomain variant V6 is confirmed more clearly in the dark-field TEM micrographs (Fig. 5.6B and C) obtained from intensity maxima of streaks corresponding to the nanodomain variants V5 and V6, respectively. The reason for the preferential growth of the nanodomain variant V6 is explained in terms of the lattice distortion strain under the stress as shown in Fig. 5.7 [23]. Cold rolling exerts compressive stress in the normal

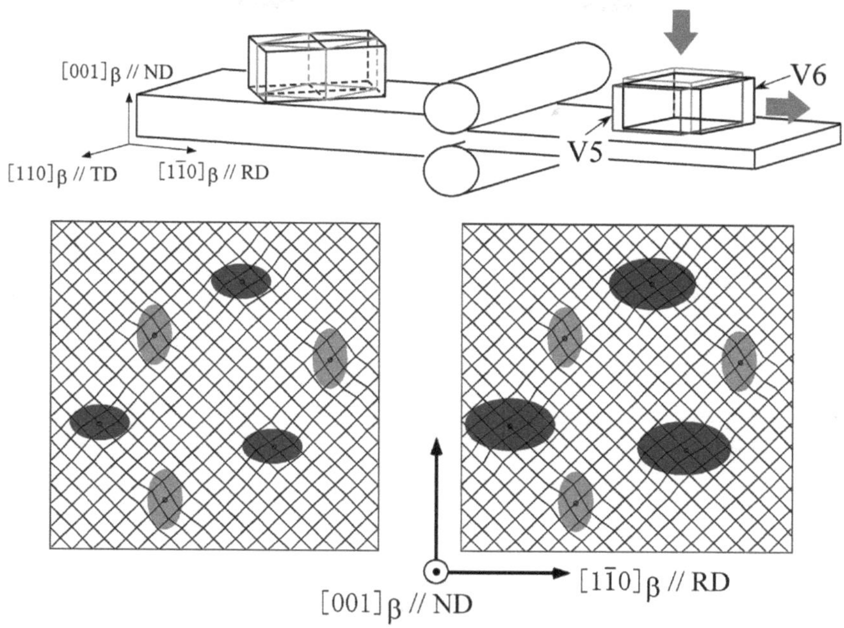

Fig. 5.7 A schematic two-dimensional diagram showing the change of nanodomain morphology as a result of cold rolling. V6 is a favorable variant upon cold rolling.

direction (ND) and tensile stress along the rolling direction (RD). This stress state causes the sample to shrink along the ND and elongate along the RD, implying that the nanodomain variant V6 is most favorable to minimize elastic energy according to the lattice correspondence between β phase and α'' phase.

Invar-like behavior along the RD can be explained by the temperature response of the favorable nanodomain variant. The nanodomains grow as the temperature decreases since the martensite phase becomes more stable. In the as-rolled specimen, the growth of nanodomains is asymmetrical: the favorably oriented nanodomains grow more easily with decreasing temperature because of the internal residual stress. Note that the first-order long-range martensitic transformation is suppressed by local stress barriers of other nanodomain variants that distributed randomly. The growth of favorable nanodomains produces a positive strain along the RD although it is difficult to measure exact magnitude of strain achieved by the growth of nanodomain because the local structure of nanodomain is unclear. As the temperature increases, the favorable nanodomains shrink gradually, causing a negative strain along the RD. It has been confirmed by XRD measurements that the lattice constant of the β phase increases almost linearly with a slope of about $8 \times 10^{-6}\ K^{-1}$, indicating that the phonon effect to thermal expansion in gum metal is similar to conventional β Ti alloys. As a result, it is proposed that the origin of anomalous low linear expansion coefficient, i.e., Invar-like behavior, in gum metals is a compensation between the positive phonon effect of lattice vibration and the negative strain component due to the lattice distortion strain of nanodomains as shown in Fig. 5.8A. This model can successfully explain not only Invar-like behavior along RD but also the anisotropic thermal expansion behavior between the RD and TD with no inconsistencies. If the lattice distortion strain effect is larger than the lattice vibration effect, negative thermal expansion is possible as shown in Fig. 5.8B.

Fig. 5.9 shows the effect of (a) oxygen content and (b) Nb content on the thermal expansion coefficient [24]. In the annealed alloys, the thermal expansion coefficient does not show significant differences due to the change of oxygen and Nb contents. However, in the as-rolled alloys,

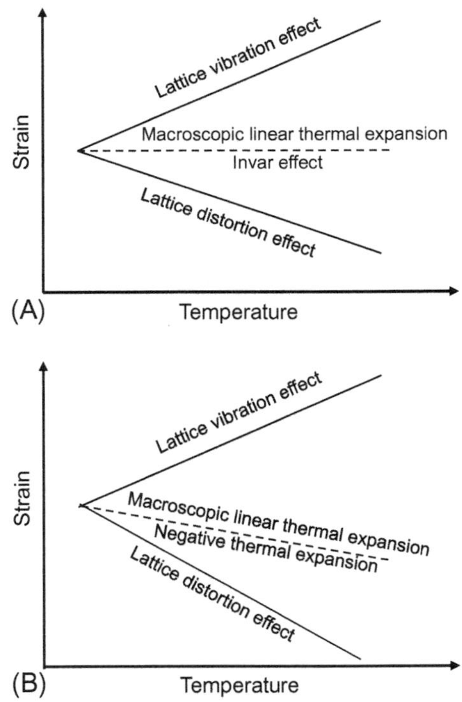

Fig. 5.8 Schematic explanations of the effect of lattice vibration and lattice distortion strain of nanodomains on the macroscopic linear thermal expansion for (A) Invar effect and (B) negative thermal expansion.

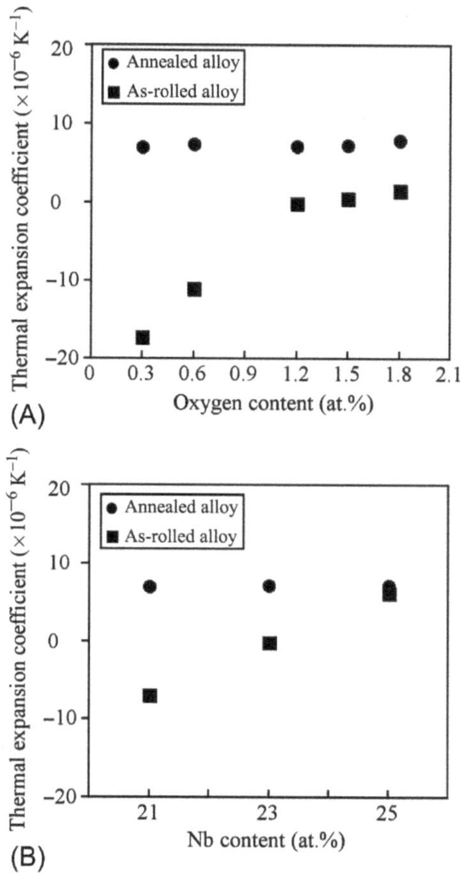

Fig. 5.9 The effect of (A) oxygen content in the Ti-23Nb-2Zr-0.7Ta-(0.3–1.8)O (at.%) alloys and (B) Nb content in the Ti-(21–25)Nb-2Zr-0.7Ta-1.2O (at.%) alloys on the thermal expansion coefficient along the RD direction.

thermal expansion coefficient is strongly influenced by not only oxygen content but also Nb content. The 0.3O alloy exhibits a large negative thermal expansion coefficient of $-18 \times 10^{-6}\ K^{-1}$. As the oxygen content increases, the thermal expansion coefficient increases and then becomes almost zero at 1.2 at.%. Further increase in the oxygen content results in only slight increase in the thermal expansion coefficient. The lattice vibration effect on the thermal expansion coefficient is found to be not significantly dependent on the oxygen content of the alloy. As a result, the oxygen content dependence of the thermal expansion coefficient is explained by the difference of the contribution of the lattice distortion. The large negative thermal expansion coefficient in the alloys with lower oxygen content indicates that the contribution of lattice distortion strain is much stronger than the lattice vibration effect because the local barrier against martensitic transformation is weak. As the oxygen content increases, the local barrier becomes strong, resulting in the small lattice distortion strain that can offset the lattice vibration effect. The Nb content also affects the lattice distortion strain. As the Nb content decreases, the stability of the β phase decreases, and then, the growth of nanodomain becomes easier, resulting in a large lattice distortion strain. As a result, the 21Nb alloy exhibits a large negative thermal expansion coefficient. On the other hand, when the Nb content is 25 at.%, the β phase is stable, and thermodynamic driving force for the growth of the nanodomain is small. Therefore, the contribution of the lattice distortion strain is negligible in the 25Nb alloy, and the thermal expansion coefficient of the alloy exhibits a positive value. These results support that the thermal expansion behavior of gum metals is controlled by the magnitude of lattice strain of nanodomains dominated by the stability of the β phase, oxygen content, and stress state.

5.3 {332} TWINNING

Deformation twinning is one of the fundamental deformation mechanism, particularly favorable in metallic materials with low-symmetry crystal structures and nanocrystalline metals [45, 46]. There have been extensive studies on the mechanism of deformation twinning and its effect on the mechanical properties [45–49]. Recently, the deformation twinning-induced plasticity (TWIP) effect has attracted attention because it can simultaneously improve both strength and ductility due to the enhanced strain hardening [50–53].

5.3.1 Deformation Modes in Metastable in β Ti Alloys

$\{332\}\langle 11\overline{3}\rangle$ twinning has received considerable attention as a unique twinning mode in metastable β Ti alloys, because the common deformation twinning mode in most of bcc metals and alloys is $\{112\}\langle\overline{1}\overline{1}1\rangle$. The $\{332\}\langle 11\overline{3}\rangle$ twinning was first identified in a Ti-11.5Mo-6Zr-4.5Sn (wt %) (Beta III) alloy deformed at room temperature in 1971 [54]. Since then, $\{332\}\langle 11\overline{3}\rangle$ twinning has been found to occur in various metastable β Ti alloys, such as Ti-V-based alloys, Ti-Cr-based alloys, Ti-Fe-based alloys, Ti-Mo-based alloys, and Ti-Nb-based alloys [55–58]. Hanada et al. [56–58] have conducted systematic research on the deformation mechanisms and mechanical properties in a series of Ti alloys. They have shown that the deformation mechanism is governed by the stability of the β phase, and $\{332\}\langle 11\bar{3}\rangle$ twinning only occurs in unstable β-type Ti alloys irrespective of the kind of alloying elements. They also claimed that the stress-induced ω-phase transformation is related to the formation of $\{332\}\langle 11\bar{3}\rangle$ twinning [58]. The deformation mode changes to dislocation slip as the stability of the β phase increases [56–58]. Effect of β-phase stability on the activity of $\{332\}\langle 11\bar{3}\rangle$ twinning has been also confirmed in the multilayered Ti-10Mo-Fe alloy consisting of Ti-10Mo-1Fe/Ti-10Mo-3Fe alternating stack [59]. The $\{332\}\langle 11\bar{3}\rangle$ twinning was found to form only in the Ti-10Mo-1Fe region where the stability of the β phase is relatively low, whereas Ti-10Mo-3Fe region deformed by dislocation slips although both regions have same crystal orientation [59]. On the other hand, it should be noted that the stress-induced martensitic transformation occurs when the β phase becomes more unstable since the onset stress for the stress-induced martensitic transformation becomes lower than the critical stress for $\{332\}\langle 11\bar{3}\rangle$ twinning. As a result, it is apparent that, as shown schematically in Fig. 5.10, the dominant deformation mode changes from stress-induced martensitic transformation to $\{332\}\langle 11\bar{3}\rangle$ twinning and then from $\{332\}\langle 11\bar{3}\rangle$ twinning to dislocation slip as the stability of β phase increases.

The interstitial alloying element of oxygen was reported to suppress the stress-induced ω phase and activate the $\{332\}\langle 11\bar{3}\rangle$ twinning in a Ti-20V alloy [60]. Deformation condition also affects the activity of deformation mechanisms. Temperature is an important factor affecting deformation mechanism. As the temperature increases, the β phase becomes more stable;

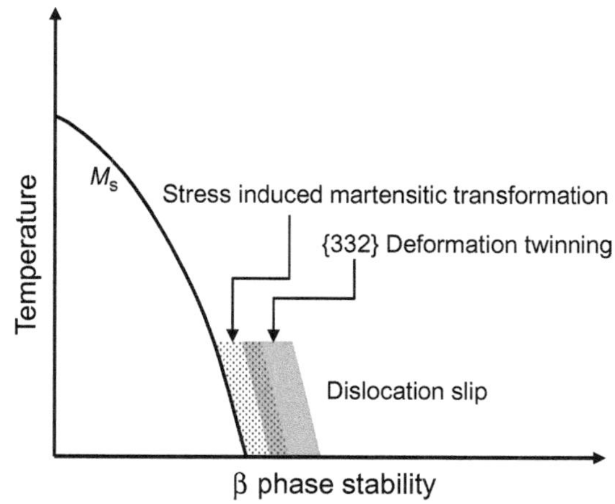

Fig. 5.10 A schematic illustration of the deformation modes in β Ti alloys.

therefore, the dominant deformation mode changes from stress-induced martensitic transformation to $\{332\}\langle 11\overline{3}\rangle$ twinning and to dislocation slip [61]. Regarding the effect of strain rate, it was reported in Ti-10V-3Fe-3Al (wt%) that the dominant deformation mechanism changes from the stress-induced martensitic transformation to $\{332\}\{332\}\langle 11\overline{3}\rangle$ twinning with increasing strain rate [62], suggesting that the increase of strain rate suppresses the martensitic transformation but assists the nucleation and propagation of $\{332\}\langle 11\overline{3}\rangle$ twinning. On the other hand, the twinning rate and morphology of $\{332\}\langle 11\overline{3}\rangle$ twinning were reported not to be dependent on the strain rate in Ti-25Nb-3Zr-3Mo-2Sn (wt%) alloy [63]. The grain orientation also affects the activity of $\{332\}\langle 11\overline{3}\rangle$ twinning [64, 65]. Hanada and Izumi [64] showed, in a Ti-15Mo-5Zr (wt%) single crystal, that the $\{332\}\langle 11\overline{3}\rangle$ twinning occurs preferentially when a tensile axis is close to the $\langle 111\rangle$ direction, whereas the dislocation slip is a dominating mechanism when the crystal is loaded near the $\langle 110\rangle$ and $\langle 001\rangle$ directions. This orientation dependence of the activity of $\{332\}\langle 11\overline{3}\rangle$ twinning is reasonably explained by Schmid's law: the $\{332\}\langle 11\overline{3}\rangle$ twinning preferentially occurs in orientations having a large Schmid factor. Similar orientation dependence has been reported in polycrystalline structures [65, 66]. Min et al. [65] reported that the area fraction of $\{332\}\langle 11\overline{3}\rangle$ twinning is strongly dependent on grain orientation in a Ti-15Mo (wt%) alloy. Bertrand et al. [66] also presented that the Schmid factor is a pertinent parameter to determine the dominating twinning system in a Ti-25Ta-24Nb (wt%) alloy.

$\{332\}\langle 11\overline{3}\rangle$ twinning plays an important role on mechanical properties of β Ti alloys. It has been reported that $\{\{332\}\langle 11\overline{3}\rangle$ twinning leads to low yield stress, high strain-hardening rate, and large elongation in many alloys including commercial alloys such as Ti-11.5Mo-6Zr-4.5Sn (wt%) and Ti-15Mo-5Zr (wt%) [26]. Recently, the $\{332\}\langle 11\overline{3}\rangle$ twinning and its contribution on mechanical properties have attracted much attention in metastable β Ti alloys. Min et al. [27, 67, 68] reported that the $\{332\}\langle 11\overline{3}\rangle$ twinning enhances the strain-hardening rate, resulting in the increased uninform elongation, in Ti-Mo-based alloys due to a dynamic grain refinement effect that decreases the dislocation mean free path. Sun et al. [69, 70] and Zhang et al. [71] reported the combined effects of twinning-induced plasticity (TWIP) and transformation-induced plasticity (TRIP) in a Ti-12Mo (wt%) alloy. They showed that multiple deformation modes, martensitic transformations (β to ω and β to α″), and twinning ($\{332\}\langle 11\overline{3}\rangle$ and $\{112\}\langle\overline{11}1\rangle$) activate simultaneously at the early stage of plastic deformation in Ti-12Mo (wt%), resulting in high strain-hardening rate, high tensile strength, and large uniform elongation. These results provided a new alloy design concept, i.e., the control of β stability in order to generate multiple deformation modes such as TRIP, TWIP, and dislocation slip, to solve the poor ductility problem of Ti alloys [72].

5.3.2 Origin of {332} Twinning

Fig. 5.11 shows the schematic illustration of twinning. Twinning modes are characterized by four elements: K_1 (twinning plane), K_2 (conjugate twinning plane), η_1 (twinning direction), and η_2 (conjugate twinning direction) [38]. In addition, the values of s and q are key factors representing the magnitude of shear and the complexity of the shuffle mechanism, respectively. The q value corresponds to the number of lattice planes parallel to the twinning plane (K_1). As shown in Fig. 5.11, all the lattice points are moved to their correct twin positions when $q=1$ or 2. On the other hand, some of the lattice points need to be rearranged after shearing to satisfy the twin relation, i.e., shuffle of atoms is required, when $q>2$. For example, half of the atoms need shuffling to form twinning symmetry when $q=4$. Table 5.1 lists possible twin systems with their twinning elements

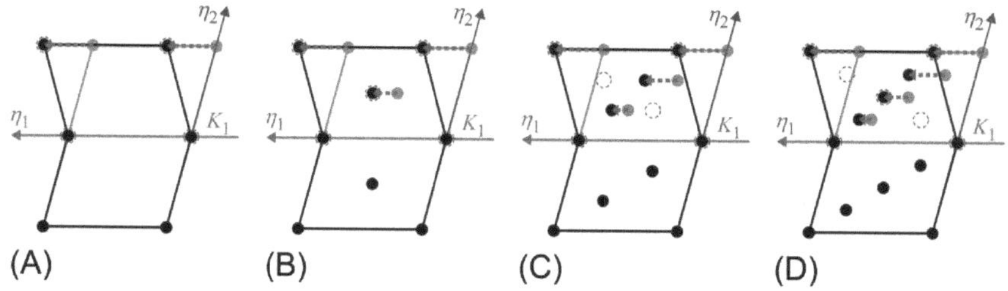

Fig. 5.11 Twinning modes with (A) $q=1$, (B) $q=2$, (C) $q=3$, and (D) $q=4$ in single-lattice structures.

Table 5.1 Twinning modes in a bcc structure

Mode	s	K_1	K_2	η_1	η_2	q
1	0.3536	{112}	$\{33\bar{2}\}$	$\langle 11\bar{1}\rangle$	$\langle 113\rangle$	4
2	0.3536	{332}	$\{11\bar{2}\}$	$\langle 11\bar{3}\rangle$	$\langle 111\rangle$	4
3	0.6124	{110}	$\{\bar{5}, 11, 8\}$	$\langle \bar{1}11\rangle$	$\langle 53\bar{1}\rangle$	4
4	0.6124	{5,11,8}	$\{\bar{1}10\}$	$\langle \bar{5}3\bar{1}\rangle$	$\langle 111\rangle$	4
5	0.7071	{112}	$\{\overline{11}2\}$	$\langle \overline{11}1\rangle$	$\langle 111\rangle$	2
6	0.7071	{110}	{114}	$\langle 001\rangle$	$\langle 22\bar{1}\rangle$	4
7	0.7071	{114}	{110}	$\langle 22\bar{1}\rangle$	$\langle 001\rangle$	4

(K_1, K_2, η_1, and η_2) and twinning shear s and q, in the bcc structure calculated by the theory of deformation twinning of Bilby and Crocker [73, 74]. $\{112\}\langle 11\bar{1}\rangle$ twinning and $\{332\}\langle 11\bar{3}\rangle$ twinning have the smallest s of 0.3536. Only $\{112\}\langle \overline{11}1\rangle$ twinning mode has $q=2$, and therefore, it is not necessary to make any rearrangement. On the other hand, other twinning modes including $\{332\}\langle 11\bar{3}\rangle$ require shuffle of atoms to form twinning symmetry. This fact explains why the $\{112\}\langle \overline{11}1\rangle$ twinning is the dominant twinning mode in the alloys with a bcc structure although the magnitude of twinning shear of this twinning mode is two times larger than the twinning modes of $\{112\}\langle 11\bar{1}\rangle$ twinning and $\{332\}\langle 11\bar{3}\rangle$.

Fig. 5.12 shows the schematic illustration of the atomic motion by the $\{332\}\langle 11\bar{3}\rangle$ twinning in the bcc structure [38]. To form the $\{332\}\langle 11\bar{3}\rangle$ twin, the shuffle of atoms is required in addition to twinning shear [75]. The magnitude of shuffle $\left(\sqrt{3/4}a_0 = 0.433a_0\right)$ for the $\{332\}\langle 11\bar{3}\rangle$ twin is quite large; therefore, it is speculated that this twinning mode is unlikely to occur. Attempts have been made to explain the atomic movement for the $\{332\}\langle 11\bar{3}\rangle$ twinning [31–34]. Kawabata et al. [31] proposed that $\{332\}\langle 11\bar{3}\rangle$ twinning progresses via a successive slip of a pair of stacked twinning dislocations with Burgers vector $1/22\langle 11\bar{3}\rangle$ on $\{332\}$ planes

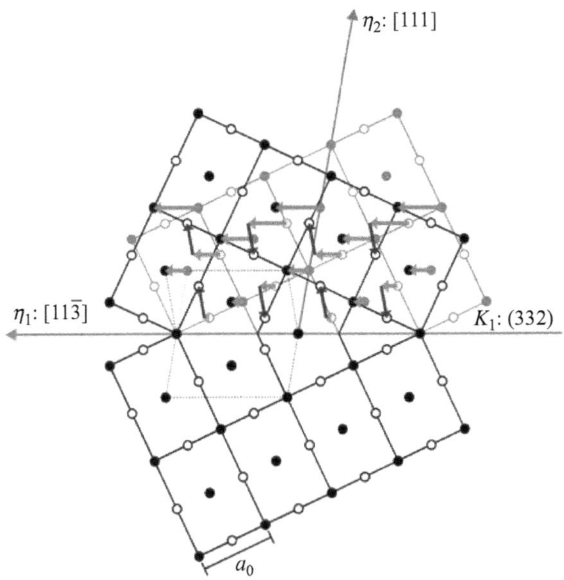

Fig. 5.12 A schematic illustration of the atomic motion by the $\{332\}\langle 11\bar{3}\rangle$ twinning in the bcc structure.

and additional shuffle by vectors $\pm 1/11\langle 332\rangle$. Similar dislocation models have been suggested by Rusakov and Litivinov [33, 34]. However, these dislocation models do not explain why $\{332\}\langle 11\overline{3}\rangle$ twinning occurs instead of $\{112\}\langle\overline{1}\overline{1}1\rangle$ twinning.

Tobe et al. [38] proposed the modulated bcc structure as shown in Fig. 5.13, in order to explain how the lattice instability facilitates the $\{332\}\langle 113\rangle$ twinning. As explained in Chapter 1, the process of β to α″ martensitic transformation can be described as follows: (a) shuffling of parallel adjacent $\{110\}$ planes to $\langle 1\overline{1}0\rangle$ directions and (b) pure shear on the $\{112\}$ planes to $\langle 11\overline{1}\rangle$ directions. They postulated that the lattice modulation, associated with shuffling of parallel adjacent $\{110\}$ planes to $\langle 1\overline{1}0\rangle$ directions, in the metastable β Ti alloys affects the twinning activities. They introduced $\delta\langle 011\rangle$ to take into account the magnitude of lattice modulation (Fig. 5.13B). The

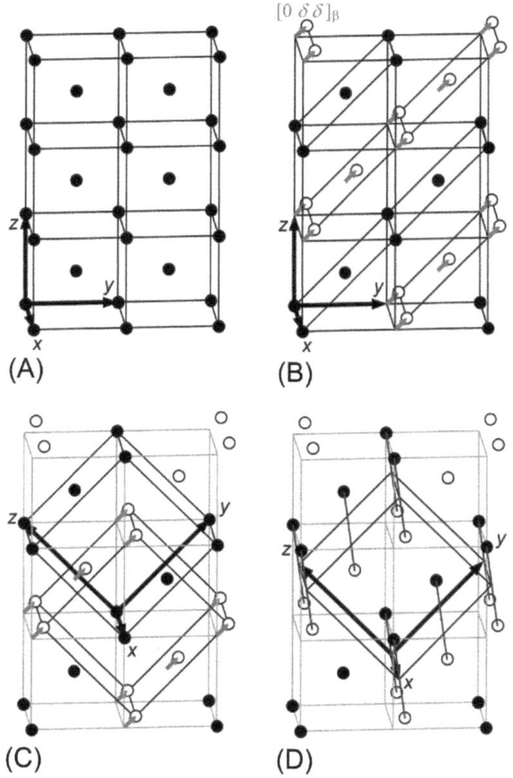

Fig. 5.13 Schematic diagram of (A) the bcc structure and (B) the modulated bcc structure. The modulated structure can be treated as the double-lattice structure with base-centered tetragonal lattices shown in (C) or (D).

modulated bcc structure can be treated as the double-lattice structure (Fig. 5.13C). And the double-lattice structure is considered as the base-centered tetragonal structure.

The $\{112\}\langle\bar{1}\bar{1}1\rangle$ and $\{332\}\langle11\bar{3}\rangle$ twinning modes in the bcc structure are corresponding to $\{110\}_t\langle1\bar{1}0\rangle_t$ and $\{130\}_t\langle\bar{3}10\rangle_t$ twinning modes in the base-centered tetragonal structure, respectively. Here, the subscript "t" refers to twinning mode in the tetragonal structure. It is noted that the shuffle of atoms, i.e., the rearrangement of motif units, is required for $\{110\}_t\langle1\bar{1}0\rangle_t$ twinning, implying that the lattice modulation makes the $\{112\}\langle\bar{1}\bar{1}1\rangle$ twinning difficult to form in the metastable β Ti alloys. On the other hand, the magnitude of shuffle of the $\{130\}_t\langle\bar{3}10\rangle_t$ twinning becomes smaller by the introduction of lattice modulation. Fig. 5.14 shows the shear and the possible shuffle mechanisms of the $\{130\}_t\langle\bar{3}10\rangle_t$ twinning when the $\delta\langle011\rangle$ is 1/24. Comparing Figs. 5.12 and 5.14, it is clear that $\{130\}_t\langle\bar{3}10\rangle_t$ twinning requires a smaller and simpler shuffle of atoms than $\{332\}\langle11\bar{3}\rangle$ twinning. It is also noted that the $\{130\}_t\langle\bar{3}10\rangle_t$ twinning is determined to be the most probable twinning mode in the tetragonal structure because it requires the smallest magnitude of shuffle when compared with other twinning modes. Furthermore, it is shown that the increase in the magnitude of $\delta\langle011\rangle$ causes enhancement of the $\{130\}_t\langle\bar{3}10\rangle_t$ twinning

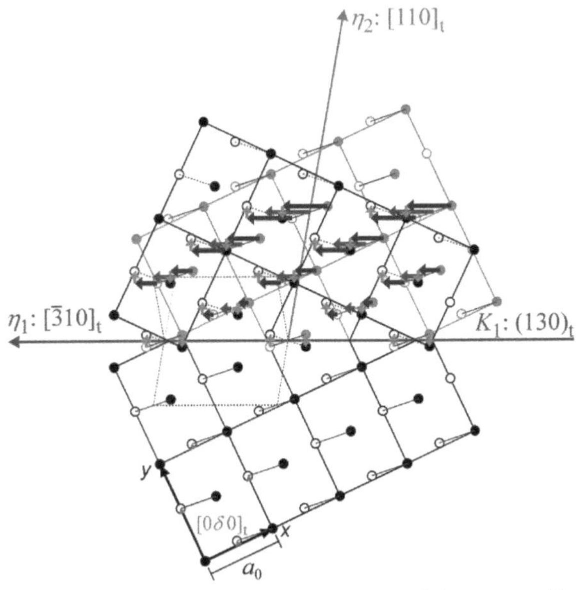

Fig. 5.14 Shear and the possible shuffle mechanisms of the $\{130\}_t\langle\bar{3}10\rangle_t$ twinning.

because the magnitude of shuffle decreases with decreasing the magnitude of $\delta\langle 011\rangle$. As a result, it is reasonably claimed that lattice modulation not only facilitates the $\{332\}\langle 11\overline{3}\rangle$ twinning but also suppresses the formation of the $\{112\}\langle\overline{1}\overline{1}1\rangle$ twinning.

Some studies have suggested that the martensitic transformation is more directly related to the formation of the $\{\{332\}\langle 11\overline{3}\rangle$ twinning [35–37, 76, 77]. Takemoto et al. [76] firstly claimed that the stress-induced martensitic transformation from β to α'' and its reverse transformation from α'' to β are involved in the formation of the $\{332\}\langle 11\overline{3}\rangle$ twinning in a Ti-4Mo (wt%) alloy. Mantani et al. [77] also presented a similar explanation involving stress-induced martensitic transformation and its reverse transformation for the formation of the $\{332\}\langle 11\overline{3}\rangle$ twinning in a Ti-40Nb (wt%). Lai et al. [35] also proposed an α''-assisted mechanism as the formation mechanism of the $\{332\}\langle 11\overline{3}\rangle$ twinning in a metastable Ti-36Nb-2Ta-3Zr (wt %) alloy by microstructural analysis using transmission electron microscopy and in situ scanning electron microscopy. They found that $\{332\}\langle 11\overline{3}\rangle$ twinning nucleates within the stress-induced α'' martensite bands and claimed that the twin band forms as a result of reverse transformation of reoriented martensite. Castany et al. [36] investigated the deformation mechanism of a superelastic Ti-27Nb (at.%) alloy, which consists of only β phase, by synchrotron x-ray diffraction (SXRD). They showed that all β phase transforms to the α'' martensite phase at the initial stage of tensile deformation. Based on microstructural analysis, they claimed that $\{332\}\langle 11\overline{3}\rangle$ twinning is not formed directly in the β phase and proposed the following stepwise scenario: (i) stress-induced martensitic transformation from β to α'', (ii) the formation of $\{130\}_{\alpha'}\langle\overline{3}10\rangle_{\alpha'}$ deformation twinning in the stress-induced α'' martensite phase, and (iii) the formation of $\{332\}\langle 11\overline{3}\rangle$ twins due to the reverse transformation from α'' to β upon unloading. It is noted that, considering the lattice correspondence between the bcc and α'' orthorhombic structures, the $\{130\}_{\alpha'}\langle\overline{3}10\rangle_{\alpha'}$ twinning in the orthorhombic phase is corresponding to that of $\{332\}\langle 11\overline{3}\rangle$ twinning in the β phase. Both models proposed by Lai et al. [35] and Castany et al. [37] are essentially the same. This mechanism is very likely to work in the alloys with relatively low β stability.

5.4 LOW YOUNG'S MODULUS

Low Young's modulus is one of the attractive features of metastable β Ti alloys. Young's modulus is closely related with the martensitic transformation in many shape memory alloys including β Ti alloys. This is because both the elastic constants and martensitic transformation temperature are closely related to the lattice stability. Particularly the shear moduli C' ($=(C_{11}-C_{12})/2$) and C_{44}, corresponding to $\{110\}\langle 1\overline{1}0\rangle$ shear modulus and $\{100\}\langle 010\rangle$ shear modulus, respectively, have been regarded as important criteria leading to martensitic transformation. Zener [78] first proposed that C' is a key parameter in determining the stability of a bcc structure: the bcc type structures become unstable as shear modulus C' decreases with decreasing temperature to martensitic transformation temperature. The softening of C' as the martensitic transformation temperature is approached has been confirmed in many shape memory alloys [79–85]. The softening of C' when approaching the martensitic transformation temperature has also been observed in β Ti alloys [13, 86]. This is reasonable because the martensitic transformation from the β phase with a bcc structure to the α″ phase with an orthorhombic structure is associated with the shear and shuffle of the $\{110\}$ plane along $\langle 1\overline{1}0\rangle$ direction.

Young's modulus of cubic crystals is determined by elastic constants, i.e., C_{11}, C_{12}, and C_{44}. Young's modulus along an arbitrary direction $\langle uvw\rangle$ in the cubic single crystals is given as

$$E_{<uvw>} = \left\{\frac{C_{11}+C_{12}}{(C_{11}-C_{12})(C_{11}+2C_{12})} + \left(\frac{1}{C_{44}} - \frac{2}{C_{11}-C_{12}}\right)\left(l^2m^2+m^2n^2+n^2l^2\right)\right\}^{-1} \tag{5.1}$$

where l, m, and n are the direction cosines of the $<uvw>$ direction and the three directions [100], [010], and [001], respectively. $E_{\langle uvw\rangle}$ can be expressed using C', C_{44}, and bulk modulus B ($=(C_{11}+2C_{12})/3$) as follows:

$$E_{<uvw>} = \left\{\left(\frac{1}{B}+\frac{3}{C'}\right) + \left(\frac{1}{C_{44}} - \frac{1}{C'}\right)\left(l^2m^2+m^2n^2+n^2l^2\right)\right\}^{-1} \tag{5.2}$$

According to Hill approximation [87], Young's modulus for the isotropic polycrystalline materials is expressed as follows:

$$E_H = \frac{1}{2}\left\{\frac{9}{1/B+15/(2C'+3C_{44})} + \frac{5}{5/9B+2/3C'+1/C_{44}}\right\} \tag{5.3}$$

Fig. 5.15 shows the Nb content dependence of Young's modulus at room temperature and martensitic transformation start temperature (M_s)

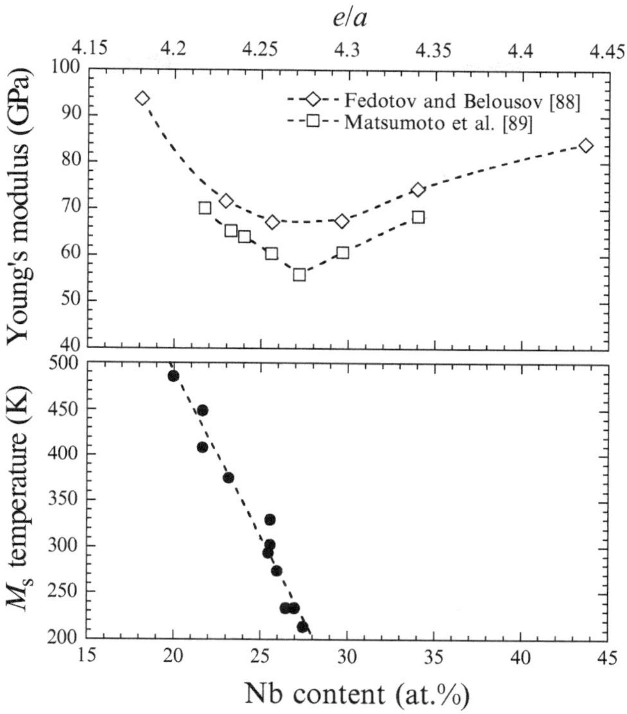

Fig. 5.15 Young's modulus at room temperature and martensitic transformation start temperature (M_s) of Ti-Nb binary alloys.

of Ti-Nb binary alloys [88–90]. It is seen that Young's modulus decreases with decreasing Nb content to a minimum reached at around 25–28 at.% and then increases with further decreasing Nb content. On the other hand, M_s decreases linearly as the Nb content increases. It is noted that Young's modulus takes a minimum value when the M_s of the alloy is near RT in both reports although the magnitudes are different. The increase in the Young's modulus with decreasing Nb content from the minimum point is due to the formation of more stable phases α'' and ω, while the increase in the Young's modulus with increasing Nb content is associated with the β-phase stability.

The single-crystal elastic constants are essential to understanding phase stability quantitatively; however, there have been limited experimental data available to date. Elastic parameters of the Ti-Nb binary alloys obtained in single crystals are listed in Table 5.2 [86, 91, 92]. For the Ti-Nb binary system, the single-crystal elastic constants of the β phase were reported in Ti-40Nb, Ti-30Nb, and Ti-25.6Nb alloys. Fig. 5.16 shows the Nb content

Table 5.2 Elastic parameters of the Ti-Nb binary alloys

Alloy	C_{11}	C_{12}	C_{44}	C'	B	A	E_h	
Ti-40Nb (at.%)	156.5	111.6	39.6	22.5	126.6	1.76	87.3	[91]
Ti-30Nb (at.%)	137.0	109.6	33.2	13.7	118.7	2.42	65.5	[92]
Ti-40Nb (wt%) (Ti-25.6Nb (at.%))	142.3	117.9	33.1	12.2	126.0	2.71	63.0	[86]

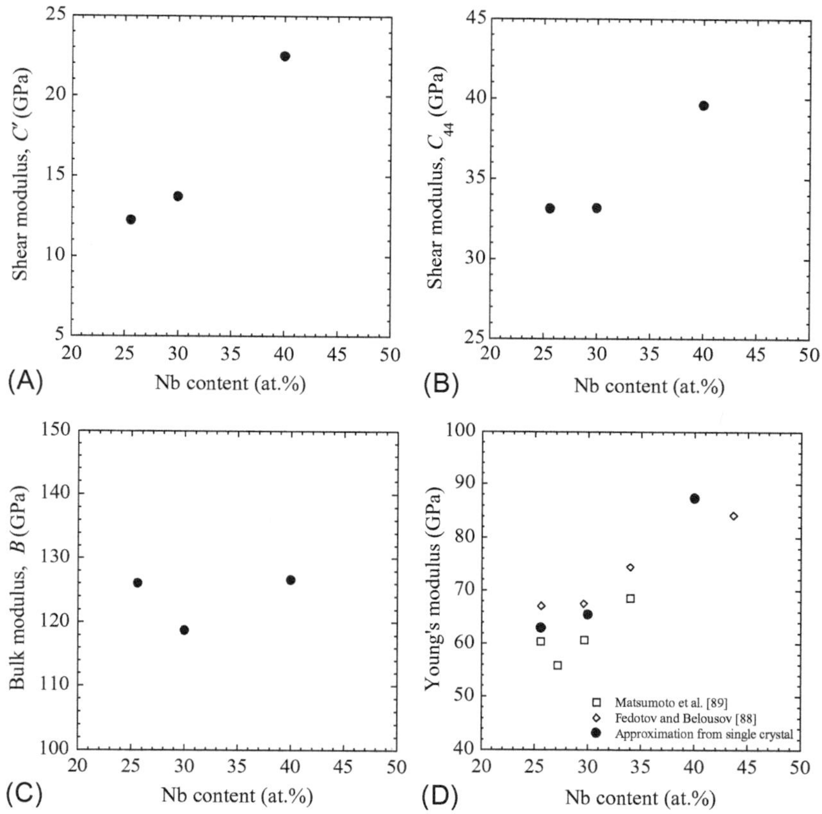

Fig. 5.16 Elastic parameters of Ti-Nb binary single crystals. (A) C', (B) C_{44}, (C) B, and (D) Young's modulus calculated by Hill approximation.

dependence of the elastic parameters: (a) C', (b) C_{44}, (c) B, and (d) Young's modulus calculated by Hill approximation (Eq. 5.3). Young's moduli measured in polycrystalline alloys are included in Fig. 5.16D. It is apparent from Fig. 5.16A that C' decreases with decreasing Nb content. The elastic constant C_{44} also decreases with decreasing Nb content although the change is not distinct when compared with C'. The bulk modulus does not show a clear dependence on the Nb content. The decrease in C' and C_{44} leads directly to decrease in Young's modulus as inferred from Eq. (5.3). Fig. 5.16D shows that the composition dependence of the measured Young's moduli in Ti-Nb binary alloys is well consistent with that of the calculated Young's moduli by Hill approximation from the elastic constants of the single crystal. From these results, it is evident that the low Young's modulus in the metastable β Ti alloys is due to the lattice softening. Particularly the

Table 5.3 Elastic parameters of the Ti-Nb-based multinary alloys

Alloy	C_{11}	C_{12}	C_{44}	C'	B	A	E_h	
Ti-29Nb-13Ta-4.6Zr (wt%)	65.1	40.5	32.4	12.3	48.7	2.63	57.2	[93]
Ti-25Nb-10Ta-5Zr (wt%)	90.8	55.9	35.8	17.5	67.5	2.05	71.1	[93]
Ti-35.7Nb-3.2Zr-2.1Ta-0.42O (wt%)	125	90	31	17.5	101.7	1.77	68.4	[44]
Ti-36.9Nb-3.0Zr-2.0Ta-0.30O (wt%)	124	93	28	15.5	103.3	1.81	61.8	[44]
Ti-29Nb-13Ta-4.6Zr (wt%)	128.4	102.4	33.1	13.0	111.1	2.32	68.9	[94]
Ti-30Nb-10Ta-5Zr (wt%)	128.3	102.1	33.0	13.1	110.8	2.52	63.9	[94]
Ti-35Nb-10Ta-5Zr (wt%)	131.0	104.2	33.2	13.4	113.1	2.48	64.8	[94]
Ti-21.9Nb-2Zr-0.7Ta-0.3O (at.%)	139.1	109.4	34.4	14.9	119.3	2.32	68.9	[13]
Ti-23.3Nb-1.9Zr-0.7Ta-1.3O (at.%)	134.5	106.5	34.1	14.0	115.8	2.44	66.9	[13]
Ti-22.4Nb-1.9Zr-0.7Ta-1.9O (at.%)	131.8	103.6	34.0	14.1	113.0	2.4	66.9	[13]
Ti-24Nb-4Zr-8Sn (wt%)	57.2	36.1	35.9	10.6	43.1	3.40	56.3	[95]
Ti-33Nb-4Sn (wt%)	126.8	101.6	22.4	12.6	110.0	1.78	50.6	[96]

decrease in the shear modulus C', which is associated with the martensitic transformation, is a key factor for low Young's modulus.

Table 5.3 shows a list of elastic parameters of the Ti-Nb multinary alloys [13, 44, 93–96]. The average number of valence electrons per atom (e/a) is a representative measure when considering the phase stability of multicomponent alloys. In order to discuss the effect of phase stability on the elastic parameters of the multinary alloys, the elastic parameters are plotted in Fig. 5.17 as a function of the e/a value of each alloy. The solid line in Fig. 5.17A is an empirical correlation between C' and e/a in bcc transition metals suggested by Fisher and Dever [97]. It is seen that C' tends to decrease as the e/a ratio decreases although there are some variations. It is also noted that the C' values of the Ti-Nb-based alloys are lower than those estimated by the empirical correlation in bcc transition metals. The calculated Young's moduli by Hill approximation also decrease with decreasing e/a ratio reflecting the decrease in C', indicating that e/a can be a dominant parameter for the design of low Young's modulus alloys.

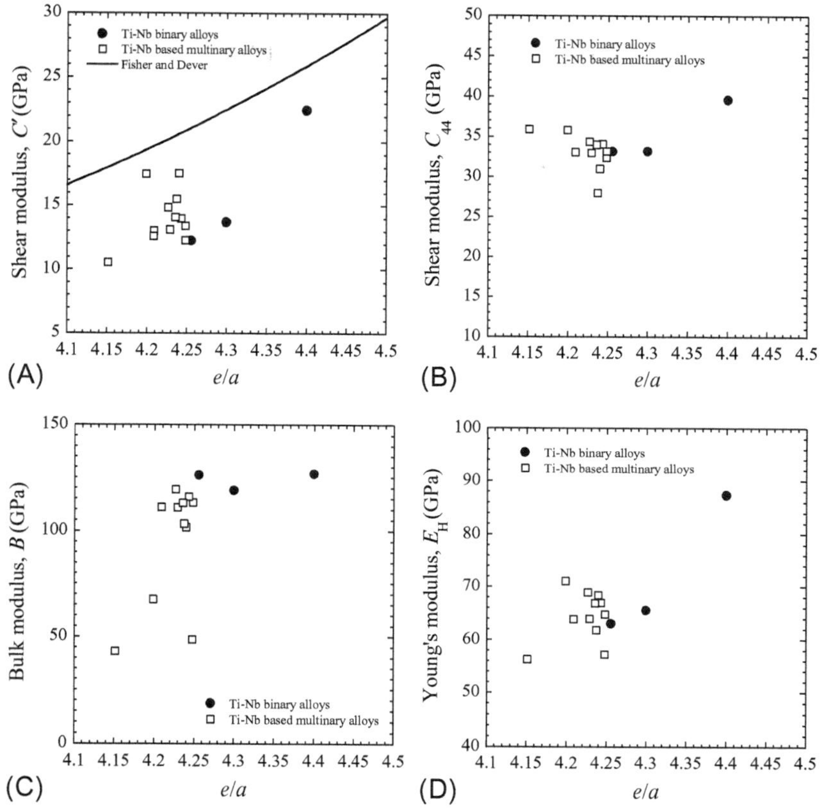

Fig. 5.17 Elastic parameters of Ti-Nb-based multinary alloys. (A) C', (B) C_{44}, (C) B, and (D) Young's modulus calculated by Hill approximation.

Table 5.4 provides a list of *e/a*, *Bo*, and *Md* of low Young's modulus alloys developed recently for biomedical applications [89, 98–105]. *Bo* and *Md* are electronic parameters calculated by the DV-Xα molecular orbital method proposed by Morinaga et al. [106] to predict the phase stability of Ti alloys. *Bo* is a bond order that is a measure of covalent bond strength between Ti and an alloying element, whereas *Md* is a d electron orbital that is related with electronegativity and metallic radius of an alloying element. Young's modulus data in Table 5.3 are plotted in Fig. 5.18 as a function of *e/a*. It is clearly seen that Young's modulus decreases with decreasing *e/a* in Ti-Nb-based multinary alloys. On the other hand, in Ti-Nb binary alloys, Young's modulus decreases with decreasing *e/a* to a minimum reached (about 60 GPa) at around 4.25–4.27 and then increases with further decreasing *e/a*. This is mainly due to the formation of the martensite phase in the

Table 5.4 List of *e/a*, *Md*, and *Bo* of low Young's modulus β Ti alloys

Alloy	*e/a*	*Md* (eV)	*Bo*	Young's modulus (GPa)	
(Ti-35Nb)-4Sn (wt%)	4.213	2.435	2.846	55	[89]
(Ti-35Nb)-7.9Sn (wt%)	4.209	2.428	2.834	62	[89]
Ti-41.1Nb-7.1Zr (wt%)	4.276	2.464	2.890	65	[98]
Ti-35.3Nb-7.1Zr-5.1Ta (wt%)	4.258	2.467	2.885	63	[99]
Ti-35Nb-7Zr-5Ta (wt%)	4.255	2.467	2.884	59.9	[100]
Ti-22.5Nb-2Zr-0.7Ta (at.%)	4.232	2.452	2.868	56	[101]
Ti-24Nb-4Zr-7.9Sn (wt%)	4.151	2.443	2.825	48.5	[102]
Ti-29Nb-6Ta-5Zr (wt%)	4.209	2.460	2.865	43	[103]
Ti-29Nb-11Ta-5Zr (wt%)	4.237	2.463	2.875	50	[103]
Ti-29Nb-13Ta-4.6Zr (wt%)	4.248	2.462	2.878	65	[104]
Ti-24Nb (at.%)	4.240	2.441	2.864	66.7	[105]
Ti-20Nb-12Zr (at.%)	4.200	2.501	2.887	54.8	[105]
Ti-17Nb-21Zr (at.%)	4.170	2.545	2.905	52.4	[105]
Ti-11Nb-38Zr (at.%)	4.110	2.630	2.936	38.8	[105]

Ti-Nb binary alloys because M_s temperature increases above RT when *e/a* decreases <4.25 as explained above. In addition, as *e/a* decreases, the athermal ω phase tends to be formed, causing the increase in Young's modulus. In multinary alloys, low Young's modulus <60 GPa can be achieved. This is due to the fact that the critical lower limit of *e/a* to maintain the β phase is decreased by the addition of alloying elements. Particularly Zr and Sn are very effective alloying elements to decrease M_s temperature while keeping low *e/a* value. This result explains why the most developed alloys for low Young's modulus contain Zr and/or Sn. It should be noted that the stress-induced martensitic transformation occurs at a low stress level when the M_s of the alloy is near RT; therefore, the martensitic transformation should be suppressed to achieve a higher yield stress. The addition of interstitial alloying elements and the introduction of plastic deformation have been found to be effective not only in suppressing martensitic transformation but also in

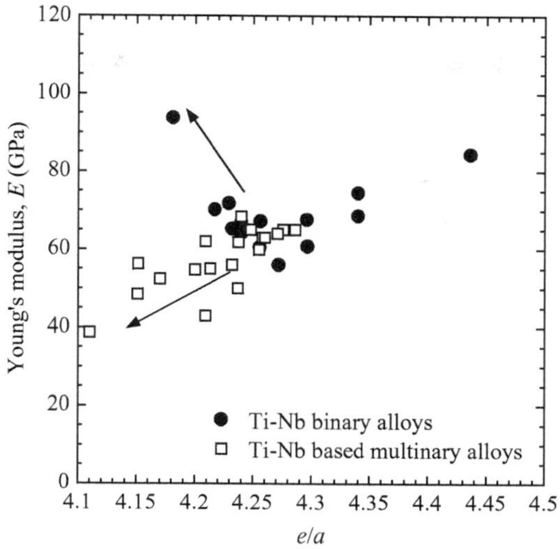

Fig. 5.18 Young's modulus of Ti-Nb polycrystalline alloys.

increasing strength of the alloy. In order to compare the electronic parameters of low Young's modulus alloys and superelastic alloys, a list of Ti-Nb-based superelastic alloys presented in Chapter 2 is shown in Table 5.5 with their *e/a*, *Bo*, and *Md*. Fig. 5.19 shows *Bo-Md* diagram of low Young's modulus alloys and superelastic alloys. It is seen that the ranges of *Bo* and *Md* values are overlapped considerably. These results are very reasonable because both properties are resultant from essentially the same mechanism, i.e., lattice softening of the β phase.

5.5 SUMMARY

Many interesting and unique characteristics of gum metals are closely related to the martensitic transformation. The mechanical properties of gum metals are strongly dependent on oxygen content. When the oxygen content is small, the alloy exhibits superelasticity due to the stress-induced martensitic transformation. As the oxygen content increases, the martensitic transformation is suppressed, and the deformation changes to the nonlinear elastic behavior with a small hysteresis. The suppression of long-range martensitic transformation is associated with the formation of multivariant nanodomain structure that is considered as local martensite-like structure. The thermal expansion behavior of gum metals is influenced by the magnitude of lattice

Table 5.5 List of *e/a*, *Md*, and *Bo* of superelastic β Ti alloys

Alloy	*e/a*	*Md* (eV)	*Bo*
Ti-26Nb	4.26	2.441	2.870
Ti-27Nb	4.27	2.441	2.873
Ti-22Nb-6Zr	4.22	2.471	2.876
Ti-22Nb-6Ta	4.28	2.447	2.879
Ti-22Nb-7Ta	4.29	2.448	2.883
Ti-24Nb-3Al	4.21	2.434	2.853
Ti-19Nb-2Pt	4.31	2.397	2.838
Ti-18Zr-15Nb	4.15	2.531	2.890
Ti-24Nb-1Mo	4.26	2.437	2.867
Ti-21Nb-2Mo	4.25	2.432	2.860
Ti-18Nb-3Mo	4.24	2.428	2.854
Ti-13Nb-4Mo	4.21	2.425	2.841
Ti-22Nb-4Zr-2Ta	4.24	2.463	2.877
Ti-16Nb-3Mo-1Sn	4.22	2.425	2.843
Ti-13.5Nb-3Mo-2Sn	4.195	2.422	2.830
Ti-15Nb-3Mo-1Sn	4.21	2.426	2.839
(Ti-Zr)-1.5Mo-3Sn	4.03	2.662	2.920
(Ti-Zr)-2Mo-2Sn	4.04	2.664	2.927
Ti-24Zr-10Nb-2Sn	4.10	2.555	2.882
Ti-18Zr-13.5Nb-1Sn	4.135	2.528	2.880
Ti-18Zr-12.5Nb-2Sn	4.125	2.525	2.872
Ti-18Zr-11Nb-3Sn	4.11	2.522	2.862
Ti-18Zr-9.5Nb-4Sn	4.095	2.519	2.852
Ti-18Zr-4.5Nb-3Sn-2Mo	4.085	2.513	2.847
Ti-18Zr-14.5Nb-1Al	4.135	2.529	2.884
Ti-18Zr-14Nb-2Al	4.12	2.527	2.879
Ti-18Zr-13.5Nb-3Al	4.105	2.524	2.874
Ti-18Zr-13Nb-4Al	4.09	2.522	2.869
Ti-24Zr-12Nb	4.12	2.561	2.898
Ti-18Nb-12Zr	4.18	2.501	2.881
Ti-19Nb-10Ta	4.29	2.451	2.884
Ti-13Nb-20Ta	4.33	2.461	2.901

strain of nanodomains. Invar-like behavior in gum metals is achieved when the negative strain component due to the lattice distortion strain of nanodomains canceled out the positive phonon effect of lattice vibration. The unique twinning mode of $\{332\}\langle 11\overline{3}\rangle$ in metastable β Ti alloys is also associated with the lattice modulation and martensitic transformation. The lattice modulation facilitates the $\{332\}\langle 11\overline{3}\rangle$ twinning by decreasing the magnitude of shuffle required for twin symmetry. Low Young's modulus

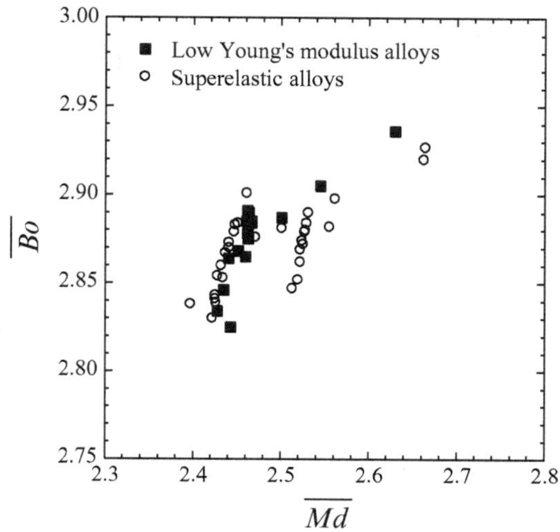

Fig. 5.19 *Bo-Md* diagram of metastable β Ti alloys showing low Young's modulus and superelasticity.

alloys and superelastic alloys have similar elastic parameters such as *e/a*, *Bo*, and *Md*, implying that both properties are related to essentially the same mechanism of lattice softening. Particularly the shear modulus *C′* is found to be a key parameter in determining phase stability.

REFERENCES

[1] T. Saito, T. Furuta, J.H. Hwang, S. Kuramoto, K. Nishino, N. Suzuki, R. Chen, A. Yamada, K. Ito, Y. Seno, T. Nonaka, H. Ikehata, N. Nagasako, C. Iwamoto, Y. Ikuhara, T. Sakuma, Multifunctional alloys obtained via a dislocation-free plastic deformation mechanism, Science 300 (2003) 464–467.
[2] T. Furuta, S. Kuramoto, J.H. Hwang, K. Nishino, T. Saito, Elastic deformation behavior of multi-functional Ti–Nb–Ta–Zr–O alloys, Mater. Trans. 46 (2005) 3001–3007.
[3] S. Kuramoto, T. Furuta, J.H. Hwang, K. Nishino, T. Saito, Elastic properties of gum metal, Mater. Sci. Eng. A 442 (2006) 454–457.
[4] S. Kuramoto, T. Furuta, J.H. Hwang, K. Nishino, T. Saito, Plastic deformation in a multifunctional Ti–Nb–Ta–Zr–O alloy, Metall. Mater. Trans. A 37A (2006) 657–662.
[5] M.Y. Gutkin, T. Ishizaki, S. Kuramoto, I.A. Ovid'ko, Nanodisturbances in deformed gum metal, Acta Mater. 54 (2006) 2489–2499.
[6] T. Li, J.W. Morris Jr, N. Nagasako, S. Kuramoto, D.C. Chrzan, Ideal engineering alloys, Phys. Rev. Lett. 98 (2007) 105503.
[7] M.Y. Gutkin, T. Ishizaki, S. Kuramoto, I.A. Ovid'ko, N.V. Skiba, Giant faults in deformed gum metal, Int. J. Plast. 24 (2008) 1333–1359.
[8] P. Castany, M. Besse, T. Gloriant, Dislocation mobility in gum metal β-titanium alloy studied via in situ transmission electron microscopy, Phys. Rev. B 84 (2011) 020201.

[9] Y. Nii, H.Y. Kim, S. Miyazaki, Effect of Nb content on plastic deformation behavior and deformation textures of Ti–Nb–Zr–Ta–O alloy, J. Japan Inst. Metals 72 (2008) 970–974.
[10] T. Yano, Y. Murakami, D. Shindo, S. Kuramoto, Study of the nanostructure of gum metal using energy-filtered transmission electron microscopy, Acta Mater. 57 (2009) 628–633.
[11] Y. Yang, S.Q. Wu, G.P. Li, Y.L. Li, Y.F. Lu, K. Yang, P. Ge, Evolution of deformation mechanisms of Ti–22.4Nb–0.73Ta–2Zr–1.34O alloy during straining, Acta Mater. 58 (2010) 2778–2787.
[12] E.A. Withey, A.M. Minor, D.C. Chrzan, J.W. Morris Jr, S. Kuramoto, The deformation of gum metal through in situ compression of nanopillars, Acta Mater. 58 (2010) 2652–2665.
[13] M. Tane, T. Nakano, S. Kuramoto, M. Hara, M. Niinomi, N. Takesue, T. Yano, H. Nakajima, Low Young's modulus in Ti–Nb–Ta–Zr–O alloys: cold working and oxygen effects, Acta Mater. 59 (2011) 6975–6988.
[14] M. Besse, P. Castany, T. Gloriant, Mechanisms of deformation in gum metal TNTZ-O and TNTZ titanium alloys: a comparative study on the oxygen influence, Acta Mater. 59 (2011) 5982–5988.
[15] M. Tane, T. Nakano, S. Kuramoto, M. Niinomi, N. Takesue, H. Nakajima, ω Transformation in cold-worked Ti–Nb–Ta–Zr–O alloys with low body-centered cubic phase stability and its correlation with their elastic properties, Acta Mater. 61 (2013) 139–150.
[16] T. Furuta, S. Kuramoto, J.W. Morris Jr, N. Nagasako, E. Withey, D.C. Chrzan, The mechanism of strength and deformation in gum metal, Scr. Mater. 68 (2013) 767–772.
[17] M.J. Lai, C.C. Tasan, D. Raabe, Deformation mechanism of ω-enriched Ti–Nb-based gum metal: dislocation channelling and deformation induced ω–β transformation, Acta Mater. 100 (2015) 290–300.
[18] Y. Yang, G.P. Li, G.M. Cheng, H. Wang, M. Zhang, F. Xu, K. Yang, Stress-introduced α″ martensite and twinning in a multifunctional titanium alloy, Scr. Mater. 58 (2008) 9–12.
[19] R.J. Talling, R.J. Dashwood, M. Jackson, D. Dye, Compositional variability in gum metal, Scr. Mater. 60 (2009) 1000–1003.
[20] R.J. Talling, R.J. Dashwood, M. Jackson, D. Dye, On the mechanism of superelasticity in gum metal, Acta Mater. 57 (2009) 1188–1198.
[21] Y. Yang, G.P. Li, G.M. Cheng, Y.L. Li, K. Yang, Multiple deformation mechanisms of Ti–22.4Nb–0.73Ta–2.0Zr–1.34O alloy, Appl. Phys. Lett. 94 (2009) 061901.
[22] J.W. Morris Jr, Y. Hanlumyuang, M. Sherburne, E. Withey, D.C. Chrzan, S. Kuramoto, Y. Hayashi, M. Hara, Anomalous transformation-induced deformation in <1 1 0> textured gum metal, Acta Mater. 58 (2010) 3271–3280.
[23] H.Y. Kim, L.S. Wei, S. Kobayashi, M. Tahara, S. Miyazaki, Nanodomain structure and its effect on abnormal thermal expansion behavior of a Ti–23Nb–2Zr–0.7Ta–1.2O alloy, Acta Mater. 61 (2013) 4874–4886.
[24] L.S. Wei, H.Y. Kim, S. Miyazaki, Effects of oxygen concentration and phase stability on nano-domain structure and thermal expansion behavior of Ti–Nb–Zr–Ta–O alloys, Acta Mater. 100 (2015) 313–322.
[25] L.S. Wei, H.Y. Kim, T. Koyano, S. Miyazaki, Effects of oxygen concentration and temperature on deformation behavior of Ti–Nb–Zr–Ta–O alloys, Scr. Mater. 123 (2016) 55–58.
[26] S. Hanada, O. Izumi, Correlation of tensile properties, deformation modes, and phase stability in commercial β-phase titanium alloys, Metall. Trans. A. 18A (1987) 265–271.
[27] X. Min, K. Tsuzaki, S. Emura, K. Tsuchiya, Enhanced uniform elongation by pre-straining with deformation twinning in high-strength β-titanium alloys with an isothermal ω-phase, Philos. Mag. Lett. 92 (2012) 726–732.

[28] M. Marteleur, F. Sun, T. Gloriant, P. Vermaut, P.J. Jacques, F. Prima, On the design of new β-metastable titanium alloys with improved work hardening rate thanks to simultaneous TRIP and TWIP effects, Scr. Mater. 66 (2012) 749–752.
[29] A.G. Crocker, Twinned martensite, Acta Metall. 10 (1962) 113–122.
[30] T. Furuhara, K. Kishimoto, T. Maki, Transmission electron microscopy of {332}<113> deformation twin in Ti–15V–3Cr–3Sn–3Al alloy, Mater. Trans. JIM 35 (1994) 843–850.
[31] T. Kawabata, S. Kawasaki, O. Izumi, Mechanical properties of TiNbTa single crystals at cryogenic temperatures, Acta Mater. 46 (1998) 2705–2715.
[32] P.G. Oberson, S. Ankem, Why twins do not grow at the speed of sound all the time, Phys. Rev. Lett. 95 (2005) 165501.
[33] V.S. Litvinov, G.M. Rusakov, Twinning on the {332}<113> system in unstable beta titanium alloys, Phys. Met. Metallogr. 90 (2000) S96–S107.
[34] G.M. Rusakov, A.V. Litvinov, V.S. Litvinov, Deformation twinning of titanium beta-alloys of transition class, Metal Sci. Heat Treat. 48 (2006) 244–251.
[35] M.J. Lai, C.C. Tasan, D. Raabe, On the mechanism of {332} twinning in metastable β titanium alloys, Acta Mater. 111 (2016) 173–186.
[36] P. Castany, Y. Yang, E. Bertrand, T. Gloriant, Reversion of a parent {130}<310> α″ martensitic twinning system at the origin of {332}<113> β twins observed in metastable β titanium alloys, Phys. Rev. Lett. 117 (2016) 245501.
[37] E. Bertrand, P. Castany, Y. Yang, E. Menou, T. Gloriant, Deformation twinning in the full-α″ martensitic Ti–25Ta–20Nb shape memory alloy, Acta Mater. 105 (2016) 94–103.
[38] H. Tobe, H.Y. Kim, T. Inamura, H. Hosoda, S. Miyazaki, Origin of {332} twinning in metastable β-Ti alloys, Acta Mater. 64 (2014) 345–355.
[39] M. Niinomi, Mechanical biocompatibilities of titanium alloys for biomedical applications, J. Mech. Behav. Biomed. Mater. 1 (2008) 30–42.
[40] M. Geetha, A.K. Singh, R. Asokamani, A.K. Gogia, Ti based biomaterials, the ultimate choice for orthopaedic implants—a review, Prog. Mater. Sci. 54 (2009) 397–425.
[41] M. Niinomi, M. Nakai, J. Hieda, Development of new metallic alloys for biomedical applications, Acta Biomater. 8 (2012) 3888–3903.
[42] M. Niinomi, Y. Liu, M. Nakai, H. Liu, H. Li, Biomedical titanium alloys with Young's moduli close to that of cortical bone, Regen. Biomater. 3 (2016) 173–185.
[43] J.G. Niu, W.T. Geng, Oxygen-induced lattice distortion in β-Ti_3Nb and its suppression effect on β to α″ transformation, Acta Mater. 81 (2014) 194–203.
[44] R.J. Talling, R.J. Dashwood, M. Jackson, S. Kuramoto, D. Dye, Determination of (C_{11} − C_{12}) in Ti–36Nb–2Ta–3Zr–0.3O (wt.%) (gum metal), Scr. Mater. 59 (2008) 669–672.
[45] J.W. Christian, S. Mahajan, Deformation twinning, Prog. Mater. Sci. 39 (1995) 1–157.
[46] Y.T. Zhu, X.Z. Liao, X.L. Wu, Deformation twinning in nanocrystalline materials, Prog. Mater. Sci. 57 (2012) 1–62.
[47] R.J. Wasilewski, On the origin of mechanical twinning—deformation or transformation? Metall. Trans. A. 8A (1977) 391–396.
[48] H. Idrissi, K. Renard, L. Ryelandt, D. Schryvers, P.J. Jacques, On the mechanism of twin formation in Fe–Mn–C TWIP steels, Acta Mater. 58 (2010) 2464–2467.
[49] E.I. Galindo-Nava, P.E.J. Rivera-Díaz-del-Castillo, Understanding martensite and twin formation in austenitic steels: a model describing TRIP and TWIP effects, Acta Mater. 128 (2017) 120–134.
[50] O. Grassel, L. Kruger, G. Frommeyer, L.W. Meyer, High strength Fe–Mn–(Al, Si) TRIP/TWIP steels development—properties—application, Int. J. Plast. 16 (2000) 1391–1409.

[51] O. Bouaziz, N. Guelton, Modeling of TWIP effect on work-hardening, Mater. Sci. Eng. A 319–321 (2001) 246–249.
[52] B.C. De Cooman, O. Kwon, K.G. Chin, State-of-the-knowledge on TWIP steel, Mater. Sci. Technol. 28 (2012) 513–527.
[53] B.C. De Cooman, Y. Estrin, S.K. Kim, Twinning-induced plasticity (TWIP) steels, Acta Mater. 142 (2018) 283–362.
[54] M.J. Blackburn, J.A. Feeney, Stress-induced transformations in Ti–Mo alloys, J. Inst. Met. 99 (1971) 132–134.
[55] M. Oka, Y. Taniguchi, {332} Deformation twins in a Ti–15.5 pct V alloy, Metall. Trans. A. 10A (1979) 651–653.
[56] S. Hanada, A. Takemura, O. Izumi, The mode of plastic deformation of β Ti–V alloys, Trans. Jpn. Inst. Metals 23 (1982) 507–517.
[57] S. Hanada, M. Ozeki, O. Izumi, Deformation characteristics in β phase Ti–Nb alloys, Metall. Trans. A. 16A (1985) 789–795.
[58] S. Hanada, O. Izumi, Transmission electron microscopic observations of mechanical twinning in metastable beta titanium alloys, Metall. Trans. A. 17A (1986) 1409–1420.
[59] X. Min, S. Emura, F. Meng, G. Mi, K. Tsuchiya, Mechanical twinning and dislocation slip multilayered deformation microstructures in β-type Ti–Mo base alloy, Scr. Mater. 102 (2015) 79–82.
[60] X.L. Wang, L. Li, H. Xing, P. Ou, J. Sun, Role of oxygen in stress-induced ω phase transformation and {3 3 2}<1 1 3> mechanical twinning in β Ti–20V alloy, Scr. Mater. 96 (2015) 37–40.
[61] H. Zhan, G. Wang, D. Kent, M. Dargusch, The dynamic response of a metastable β Ti–Nb alloy to high strain rates at room and elevated temperatures, Acta Mater. 105 (2016) 104–113.
[62] M. Ahmed, D. Wexler, G. Casillas, D.G. Savvakin, E.V. Pereloma, Strain rate dependence of deformation-induced transformation and twinning in a metastable titanium alloy, Acta Mater. 104 (2016) 190–200.
[63] H. Zhan, W. Zeng, G. Wang, D. Kent, M. Dargusch, On the deformation mechanisms and strain rate sensitivity of a metastable β Ti–Nb alloy, Scr. Mater. 107 (2015) 34–37.
[64] S. Hanada, O. Izumi, Deformation of metastable beta Ti–15Mo–5Zr alloy single crystals, Metall. Trans. A. 11A (1980) 1447–1452.
[65] X. Min, S. Emura, X. Chen, X. Zhou, K. Tsuzaki, K. Tsuchiya, Deformation microstructural evolution and strain hardening of differently oriented grains in twinning-induced plasticity β titanium alloy, Mater. Sci. Eng. A 659 (2016) 1–11.
[66] E. Bertrand, P. Castany, I. Peron, T. Gloriant, Twinning system selection in a metastable β-titanium alloy by Schmid factor analysis, Scr. Mater. 64 (2011) 1110–1113.
[67] X. Min, K. Tsuzaki, S. Emura, K. Tsuchiya, Enhancement of uniform elongation in high strength Ti–Mo based alloys by combination of deformation modes, Mater. Sci. Eng. A 528 (2011) 4569–4578.
[68] X. Min, X. Chen, S. Emura, K. Tsuchiya, Mechanism of twinning-induced plasticity in β-type Ti–15Mo alloy, Scr. Mater. 69 (2013) 393–396.
[69] F. Sun, J.Y. Zhang, M. Marteleur, T. Gloriant, P. Vermaut, D. Laille, P. Castany, C. Curfs, P.J. Jacques, F. Prima, Investigation of early stage deformation mechanisms in a metastable β titanium alloy showing combined twinning-induced plasticity and transformation-induced plasticity effects, Acta Mater. 61 (2013) 6406–6417.
[70] F. Sun, J.Y. Zhang, M. Marteleur, C. Brozek, E.F. Rauch, M. Veron, P. Vermaut, P. Jacques, F. Prima, A new titanium alloy with a combination of high strength, high strain hardening and improved ductility, Scr. Mater. 94 (2015) 17–20.

[71] J.Y. Zhang, J.S. Li, Z. Chen, Q.K. Meng, F. Sun, B.L. Shen, Microstructural evolution of a ductile metastable β titanium alloy with combined TRIP/TWIP effects, J. Alloys Compd. 699 (2017) 775–782.
[72] C. Brozek, F. Sun, P. Vermaut, Y. Millet, A. Lenain, D. Embury, P.J. Jacques, F. Prima, A β-titanium alloy with extra high strain-hardening rate: design and mechanical properties, Scr. Mater. 114 (2016) 60–64.
[73] B.A. Bilby, A.G. Crocker, The theory of the crystallography of deformation twinning, Proc. Roy. Soc. A 288 (1965) 240–255.
[74] A.G. Crocker, The crystallography of deformation twinning in alpha-uranium, J. Nucl. Mater. 16 (1965) 306–326.
[75] R.H. Richman, in: R.E. Reed-Hill et al., (Eds.), Deformation Twinning, Gordon and Breach, New York, 1964, pp. 237–271.
[76] Y. Takemoto, M. Hida, A. Sakakibara, Martensitic {332}<113> twin in β type Ti–Mo alloy, J. Japan Inst. Metals 53 (1989) 1004–1012.
[77] Y. Mantani, Y. Takemoto, M. Hida, A. Sakakibara, Formation of α″ martensite and {332}<113> twin during tensile deformation in Ti–40 mass%Nb alloy, J. Japan Inst. Metals 66 (2002) 1022–1029.
[78] C. Zener, Contributions to the theory of beta-phase alloys, Phys. Rev. 71 (1947) 846–851.
[79] N. Nakanishi, Elastic constants as they relate to lattice properties and martensite formation, Prog. Mater. Sci. 24 (1980) 143–265.
[80] B. Verliden, L. Delaey, On the elastic constants and Ms-temperatures in β-hume-rothery alloys, Acta Metall. 36 (1988) 1771–1779.
[81] S. Muto, R. Oshima, F.E. Fujita, Elastic softening and elastic strain energy consideration in the fcc-fct transformation of FePd alloys, Acta Metall. Mater. 38 (1990) 685–694.
[82] X. Ren, N. Miura, J. Zhang, K. Otsuka, K. Tanaka, M. Koiwa, T. Suzuki, Y.I. Chumlyakov, M. Asai, A comparative study of elastic constants of Ti-Ni based alloys prior to martensitic transformation, Mater. Sci. Eng. A 312 (2001) 196–206.
[83] P. Sedlak, H. Seiner, M. Landa, V. Novak, P. Sittner, L.I. Manosa, Elastic constants of bcc austenite and 2H orthorhombic martensite in CuAlNi shape memory alloy, Acta Mater. 53 (2005) 3643–3661.
[84] T. Fukuda, T. Kakeshita, Lattice softening in Fe_3Pt exhibiting three types of martensitic transformations, Metals 7 (2016) 156.
[85] K. Otsuka, X. Ren, Physical metallurgy of Ti-Ni-based shape memory alloys, Prog. Mater. Sci. 50 (2005) 511–678.
[86] H.W. Jeong, Y.S. Yoo, Y.T. Lee, J.K. Park, Elastic softening behavior of Ti–Nb single crystal near martensitic transformation temperature, J. Appl. Phys. 108 (2010) 063515.
[87] R. Hill, The elastic behaviour of a crystalline aggregate, Proc. Phys. Soc. A 65 (1952) 349–355.
[88] S.G. Fedotov, P.K. Belousov, Elastic constants of alloys of the system titanium–niobium, Phys. Met. Metallogr. 17 (1964) 83–86.
[89] H. Matsumoto, S. Watanabe, S. Hanada, Beta TiNbSn alloys with low Young's modulus and high strength, Mater. Trans. 46 (2005) 1070–1078.
[90] H.Y. Kim, Y. Ikehara, J.I. Kim, H. Hosoda, S. Miyazaki, Martensitic transformation, shape memory effect and superelasticity of Ti–Nb binary alloys, Acta Mater. 54 (2006) 2419–2429.
[91] C.N. Reid, J.L. Routbort, R.A. Maynard, Elastic constants of Ti–40 at.% Nb at 298 ° K, J. Appl. Phys. 44 (1973) 1398–1399.
[92] R. Hermann, H. Hermann, M. Calin, B. Buchnera, J. Eckert, Elastic constant of single crystalline β-$Ti_{70}Nb_{30}$, Scr. Mater. 66 (2012) 198–201.

[93] M. Tane, S. Akita, T. Nakano, K. Hagihara, Y. Umakoshi, M. Niinomi, H. Nakajima, Peculiar elastic behavior of Ti–Nb–Ta–Zr single crystals, Acta Mater. 56 (2008) 2856–2863.
[94] M. Tane a, S. Akita, T. Nakano, K. Hagihara, Y. Umakoshi, M. Niinomi, H. Mori, H. Nakajima, Low Young's modulus of Ti–Nb–Ta–Zr alloys caused by softening in shear moduli c′ and c44 near lower limit of body-centered cubic phase stability, Acta Mater. 58 (2010) 6790–6798.
[95] Y.W. Zhang, S.J. Li, E.G. Obbard, H. Wang, S.C. Wang, Y.L. Hao, R. Yang, Elastic properties of Ti–24Nb–4Zr–8Sn single crystals with bcc crystal structure, Acta Mater. 59 (2011) 3081–3090.
[96] Y.P. Hou, S. Guoa, X.L. Qiao, T. Tian, Q.K. Meng, X.N. Cheng, X.Q. Zhao, Origin of ultralow Young's modulus in a metastable β-type Ti–33Nb–4Sn alloy, J. Mech. Behav. Biomed. Mater. 59 (2016) 220–225.
[97] E.S. Fisher, D. Dever, Relation of the c′ elastic modulus to stability of b.c.c. transition metals, Acta Metall. 18 (1970) 265–269.
[98] S. Schneider, S.G. Schneider, H.M. Silva, C.M. Neto, Study of the non-linear stress–strain behavior in Ti–Nb–Zr alloys, Mater. Res. 8 (2005) 435–438.
[99] L.M. Elias, S.G. Schneider, S. Schneider, H.M. Silva, F. Malvisi, Microstructural and mechanical characterization of biomedical Ti–Nb–Zr(–Ta) alloys, Mater. Sci. Eng. A 432 (2006) 108–112.
[100] D. Raabe, B. Sander, M. Friak, D. Ma, J. Neugebauer, Theory-guided bottom-up design of β-titanium alloys as biomaterials based on first principles calculations: theory and experiments, Acta Mater. 55 (2007) 4475–4487.
[101] Q. Wei, L. Wang, Y. Fu, J. Qin, W. Lu, D. Zhang, Influence of oxygen content on microstructure and mechanical properties of Ti–Nb–Ta–Zr alloy, Mater. Des. 32 (2011) 2934–2939.
[102] Y.L. Hao, S.J. Li, S.Y. Sun, C.Y. Zheng, R. Yang, Elastic deformation behavior of Ti–24Nb–4Zr–7.9Sn for biomedical applications, Acta Biomater. 3 (2007) 277–286.
[103] P. Laheurte, F. Prima, A. Eberhardt, T. Gloriant, M. Wary, E. Patoor, Mechanical properties of low modulus β titanium alloys designed from the electronic approach, J. Mech. Behav. Biomed. Mater. 3 (2010) 565–573.
[104] M. Niinomi, T. Akahori, S. Katsura, K. Yamauchi, M. Ogawa, Mechanical characteristics and microstructure of drawn wire of Ti–29Nb–13Ta–4.6Zr for biomedical applications, Mater. Sci. Eng. C 27 (2007) 154–161.
[105] L. You, X. Song, A study of low Young's modulus Ti–Nb–Zr alloys using d electrons alloy theory, Scr. Mater. 67 (2012) 57–60.
[106] M. Morinaga, N. Yukawa, T. Maya, K. Sone, H. Adachi, in: P. Lacombe, R. Tricot, G. Beranger (Eds.), Theoretical design of titanium alloys, Proceedings of Sixth World Conference on Titanium, Societe Francaise de Metallurgie, Cannes, 1988, pp. 1601–1606.

CHAPTER 6

Biocompatibility of Superelastic Beta Ti Alloys

Contents

6.1 INTRODUCTION

Among the alloys exhibiting shape memory effect and superelasticity, only Ti-Ni-based shape memory alloys have become commercially popular due to their multiple excellent properties such as high strength, good corrosion resistance, and stable shape memory properties with a large recovery strain [1, 2]. Ti-Ni-based shape memory alloys have been successfully used as biomedical devices and materials such as stents, guide wires, and orthodontic arch wires [2, 3]. Many studies have shown that Ti-Ni shape memory alloys exhibit excellent biocompatibility and corrosion resistance [3]; however, their high Ni concentration has caused some concerns on biocompatibility since Ni can lead to deleterious effects on human health, e.g., genotoxicity, carcinogenesis, and hypersensitivity [4–10]. From this point of view, there have been extensive efforts to develop more biocompatible shape memory alloys, and many new Ti-based shape memory alloys consisting only biocompatible elements have been developed. In this chapter, the recent experimental results on cytocompatibility of Ni-free Ti-based superelastic alloys are overviewed and compared with Ti-Ni, Ti, and Ti-6Al-4V. Electrochemical properties and corrosion behavior of Ti-based superelastic alloys are also summarized because cytocompatibility is strongly related with the corrosion resistance and ion release.

Ni-free Ti-based Shape Memory Alloys
https://doi.org/10.1016/B978-0-12-809401-3.00006-9

6.2 CYTOCOMPATIBILITY

Cytocompatibility is one of the essential evaluation criteria for assessing biological safety of the materials and devices. A direct measurement of cell growth on the surface of metallic material is one of the useful methods to evaluate cytocompatibility. The growth rate and viability of cells are dependent on not only the composition of metals but also the type of cells. L929 mouse fibroblast cell is one of the recommended cell lines for in vitro cytotoxicity test by ISO10993 [11]. However, it has been reported that fibroblast cells are less sensitive for the material difference [12]. Considering the implant applications and cardiovascular applications such as fracture fixations, joint replacements, and stents, it is importance to use the cell lines from the tissue adjacent to an implanting place for the cytocompatibility evaluation.

Cytocompatibility has been evaluated for several Ni-free Ti-based superelastic alloys, including Ti-18Zr-11Nb-3Sn, Ti-24Nb-3Al, Ti-26Nb, Ti-25Ta-25Nb, and Ti-23Nb-0.7Ta-2Zr-0.5N, using different cell lines in comparison with those of pure Ti, Ti-6Al-4V, or conventional Ti-Ni shape memory alloys. Fu et al. [13] examined cytocompatibility of the Ti-18Zr-11Nb-3Sn superelastic alloy using three types of cell lines, i.e., murine fibroblast L929, human osteosarcoma (SaOS-2), and human umbilical vein endothelial cell (HUVEC). A commercial Ti-50.5Ni alloy was also investigated for comparison. The specimen of each alloy was sterilized by autoclaving at 393 K for 30 min and placed on the bottom of the 24-well microplate. Culturing condition was as follows: 800 cells in 1.2 mL of *E*-MEM + 10% FBS for L929, 10,000 cells in 1.2 mL of D-MEM + 10% FBS for SaOS-2, and 3000 cells in 1.2 mL of HuMedia-EG2 for HUVEC. After the incubation of 1, 4, and 7 days in the CO2 incubator, WST-1 assay, Giemsa staining, and cell morphology observation were performed. The results confirmed that the Ti-18Zr-11Nb-3Sn alloy has better cytocompatibility than the conventional Ti-50.5Ni alloy. Fig. 6.1 shows the comparison of cell covering ratios of the Ti-18Zr-11Nb-3Sn alloy and the Ti-50.5Ni alloy [13]. During incubation, the cells grow continuously, and the cell covering ratio increases with time. After 7 days of incubation, Ti-18Zr-11Nb-3Sn exhibited good cell growth and larger cell covering ratios for three kinds of cells when compared with those of the Ti-50.5Ni alloy, indicating a good advantage and possibility for implant application.

Kanetaka et al. [14] investigated cytocompatibility of Ti-24Nb-3Al and Ti-7Cr-3Sn superelastic alloys in comparison with commercial pure titanium

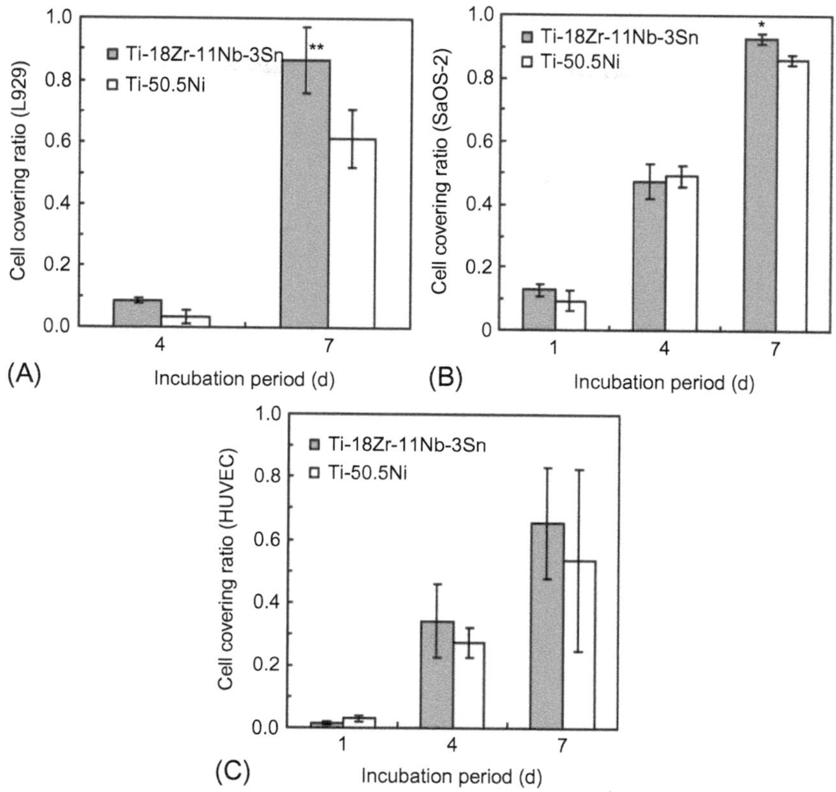

Fig. 6.1 Cell covering ratio of (A) L929, (B) SaOS-2, and (C) HUVEC onto Ti-18Zr-11Nb-3Sn and Ti-50.5Ni (mean ± standard deviation) [13]. $*p < 0.1$, $**p < 0.05$.

(cp-Ti) using African green monkey kidney fibroblast cell line COS7. They evaluated the cell proliferation and viability of the COS7 cells on the alloys by DNA quantification, live-cell imaging, and cytoskeletal observations. Fig. 6.2 shows the comparison of DNA concentration after incubation for 1, 3, and 5 days [14]. The concentration of DNA increased significantly between 1 and 3 days, while no significant change occurred between 3 and 5 days for all alloys. It is noted that the cells on Ti-24Nb-3Al and Ti show similar level of DNA concentration while the concentration of DNA from the cells on Ti-7Cr-3Sn was less than that from the cells on Ti. Also, the COS7 cells on Ti-24Nb-3Al and Ti show clearer morphology with larger size than those on Ti-Cr-Sn in fluorescent microcopy, indicating that Ti-24Nb-3Al and Ti show higher cell viability when compared with Ti-Cr-Sn. Accordingly, they concluded that Ti-Nb-Al has good cytocompatibility comparable with that of

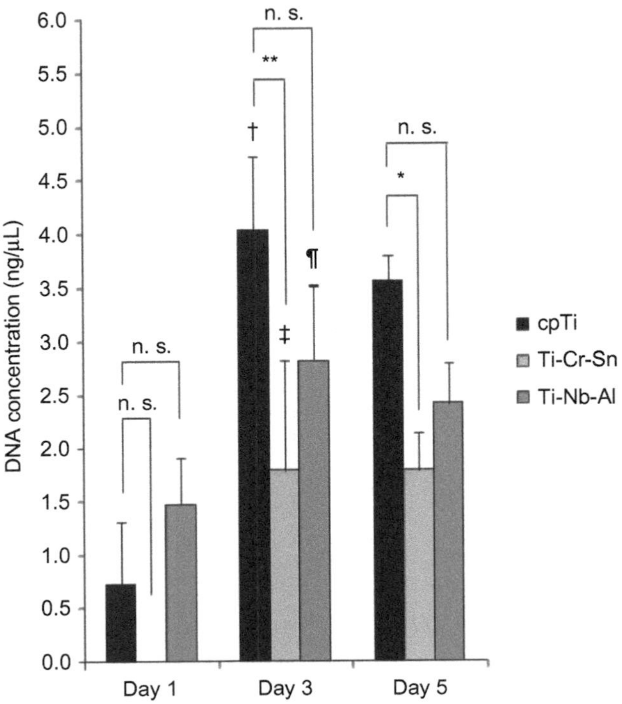

Fig. 6.2 DNA concentration in cells onto cp-Ti, Ti-Cr-Sn, and Ti-Nb-Al incubated for 1, 3, or 5 days [14]. $^{**}p < 0.01$, $^{*}p < 0.05$, $^{\dagger}p < 0.01$ vs. Ti Day 1, $^{\ddagger}P < 0.01$ vs. Ti-Cr-Sn Day 1, $^{\P}p < 0.05$ vs. Ti-Nb-Al Day 1, n.s.: not significant.

pure Ti. The relatively poor cytocompatibility in Ti-Cr-Sn is suggested to be due to the Cr^{3+} and Sn^{2+} that show relatively lower cytocompatibility than Nb^{5+}, Al^{3+}, and Sn^{4+} [5, 12].

McMahon et al. [15] reported a comparative study on cytocompatibility of Ti-26Nb and Ni-49.2Ti (Ti-50.8Ni) shape memory alloys and pure Ti. The cytotoxicity of the alloys was evaluated using NIH 3T3 mouse fibroblast cells encapsulated within 3-D PEGDA hydrogels. Fig. 6.3 shows a schematic of the testing setup (a) and the results of cell death (b) and DNA concentration (c) [15]. It is seen that the relative cell death ratio of Ni-49.2Ti is higher than those of Ti-26Nb and pure Ti after 14 days, indicating that the cytocompatibility of Ti-26Nb is comparable with that of pure Ti and superior to that of the conventional Ti-Ni shape memory alloy. The DNA concentration (Fig. 6.3C) also exhibited a similar tendency: both Ti-26Nb and pure Ti showed a higher DNA concentration in the hydrogels when compared with the conventional Ti-Ni shape memory alloy. They claimed that the superior

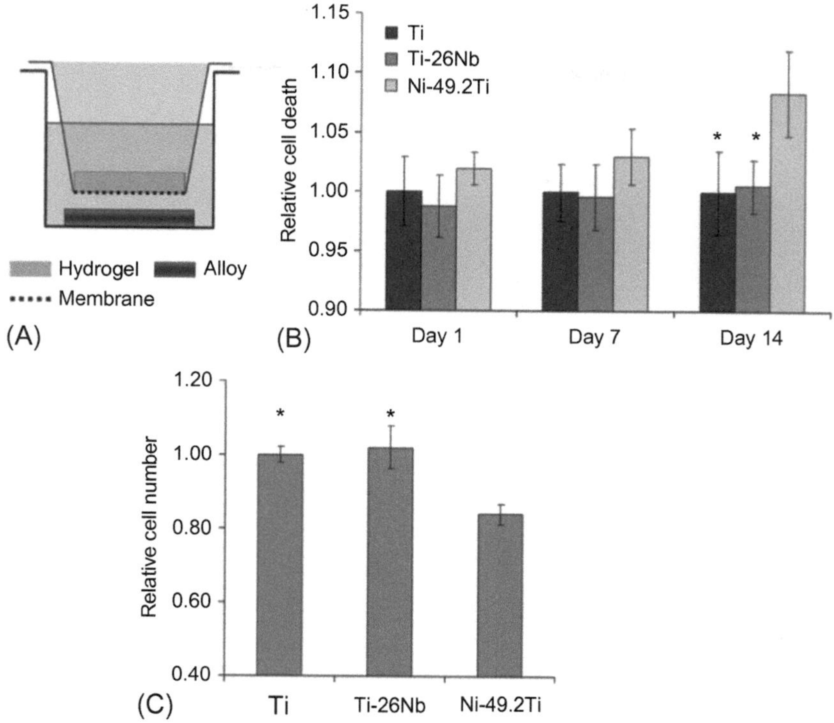

Fig. 6.3 (A) Schematic of the cytotoxicity testing setup. Cytotoxicity of Ti-26Nb alloys compared with Ni-49.2Ti and pure Ti samples by (B) LDH assay and (C) day-14 DNA assessment [15]. *$p < 0.05$.

cytocompatibility of Ti-26Nb and pure Ti is related to the increased corrosion resistance and the reduced metallic ion release.

Cimpean et al. [16] investigated the cytocompatibility of a Ti-25Ta-25Nb superelastic alloy using a human fetal osteoblast cell line (hFOB 1.19) in comparison with Ti-6Al-4V and tissue culture polystyrene (TCPS) as reference substrates. They showed that Ti-25Ta-25Nb has almost identical level of cytocompatibility with Ti-6Al-4V and TCPS in terms of viability, cell attachment and spreading, morphology of cells, and cell proliferation potential. Ion et al. [17] evaluated the cytocompatibility of a Ti-23Nb-0.7Ta-2Zr-0.5N superelastic alloy using two cell lines, i.e., MC3T3-E1 preosteoblasts and RAW 264.7 macrophages in comparison with Ti-6Al-4V as a control substrate. They reported that Ti-23Nb-0.7Ta-2Zr-0.5N shows stronger osteoblastic differentiation and mineralization ability when compared with Ti-6Al-4V. It is clearly seen in Fig. 6.4 that

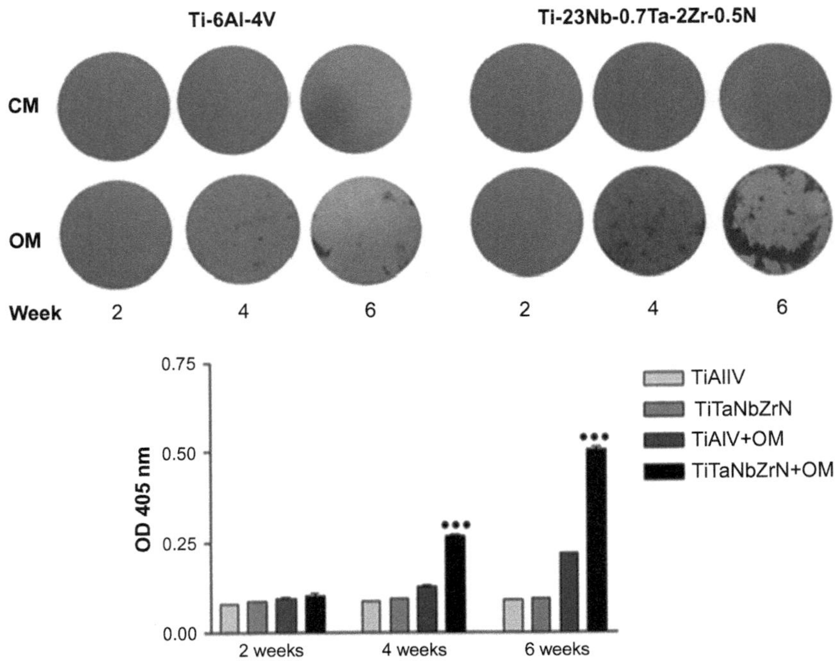

Fig. 6.4 Digital images of the analyzed samples showing extracellular matrix mineralization (upper panel) and colorimetrically quantitative analysis (lower panel) after 2, 4, and 6 weeks of incubation [17]. $^{***}p < 0.001$ compared with the Ti-6Al-4V samples maintained in osteogenic induction medium.

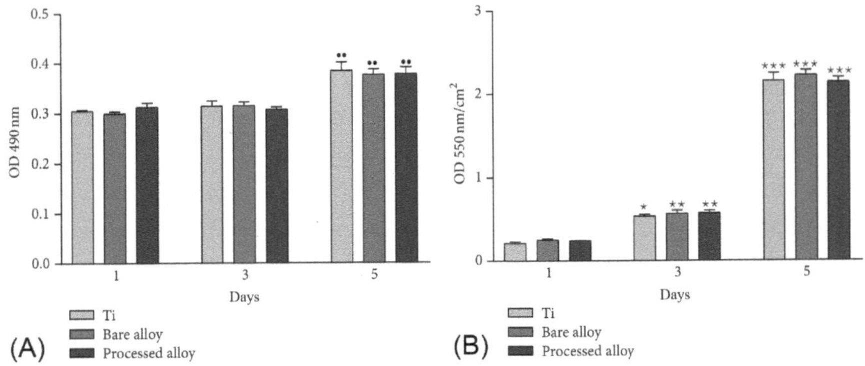

Fig. 6.5 Viability of MG63 osteoblast-like cells cultured onto Ti and bare and nitrided Ti-23Nb-0.7Ta-2Zr-0.5N alloy for 1, 3, and 5 days as determined by (A) LDH and (B) MTT assays [18]. ••$p < 0.01$ vs. corresponding sample at 1 and 3 days, $^{*}p < 0.05$ vs. corresponding sample at 1 day, $^{**}p < 0.01$ vs. corresponding sample at 1 day, $^{***}p < 0.001$ vs. corresponding sample at 1 day.

Ti-23Nb-0.7Ta-2Zr-0.5N exhibits significantly stronger mineralization ability particularly at 6 weeks of culture [17].

Mitran et al. [18, 19] also reported that Ti-23Nb-0.7Ta-2Zr-0.5N exhibits no significant cytotoxic responses on MG63 human osteoblast-like cells. Fig. 6.5 shows the comparison of viability of MG63 cells on Ti, bare Ti-23Nb-0.7Ta-2Zr-0.5N, and nitrided Ti-23Nb-0.7Ta-2Zr-0.5N determined by lactate dehydrogenase (LDH) release assay and MTT, 3-(4,5-dimethylthiazol-2-yl)-2,5-diphenyltetrazolium bromide, assay [18]. LDH release assay is a good measure for the toxic potential, and MTT represents a cell proliferation rate. Fig. 6.5A shows that both alloys exhibit similar LDH activities, i.e., similar optical density (OD) values, at any day, indicating that MG63 cells show a similar viability on Ti and Ti-23Nb-0.7Ta-2Zr-0.5N. Fig. 6.5B shows that the number of cells increased continuously with the increase of culture time with no significant differences between substrates. From the results for both assays, it is clear that the cytocompatibility of Ti-23Nb-0.7Ta-2Zr-0.5N is equal to that of pure Ti.

Ti-Nb-Sn alloys also exhibit a good cytocompatibility comparable with pure Ti. Miura et al. [20] reported that there were no significant differences in cell viability and proliferation behavior for L929 and MC3T3-E1 cell cultures in the Ti-Nb-Sn alloy, Ti-6Al-4V, and Ti. Similar results have been reported for Ti-Nb-Zr alloys [21, 22]. Arciniegas et al. [21, 22] reported that the cell adhesion response of MG63 cells in Ti-19.1Nb-9.8Zr (wt%) and Ti-41.2Nb-6.1Zr (wt%) alloys is equivalent to that in the cp-Ti. Both alloys also showed similar cell survival percentages compared with the negative control (tissue culture polystyrene (TCPS)). However, at longer culture time, Ti-19.1Nb-9.8Zr (wt%) revealed a notably lower cell proliferation rate when compared with Ti-41.2Nb-6.1Zr (wt%) and pure Ti. They claimed that the amorphous layer of Nb_2O_5 has a favorable effect on the cell adhesion and morphology response. Xue et al. [23] assessed that the hemolysis rate of Ti-19Zr-10Nb-1Fe is much lower than that of TiNi, although both alloys showed similar cytocompatibility for MC3T3-E1 osteoblast cell line. They also claimed that the Ti-19Zr-10Nb-1Fe alloy showed more stable platelet adhesion of blood when compared with TiNi.

A Ti-3Mo-6Sn-5.2Zr superelastic alloy has been found to have excellent biocompatibility equivalent to pure Ti in terms of cell proliferation rate, cell metabolic rate, and cell cycle for V-79 Chinese hamster fibroblastic cells and MC3T3-E1 cells [24]. Ion et al. [25] investigated the cell viability and cell proliferation behavior of HUVEC on a Ti-23Hf-3Mo-4Sn superelastic alloy. The results exhibited that Ti-23Hf-3Mo-4Sn has a similar level of

Table 6.1 Comparison of cytocompatibility of Ti-based superelastic alloys in comparison with control materials (Ti-Ni, pure Ti, and Ti-6Al-4V)

		Cytocompatibility of Ti-based SE alloys in comparison with control			
Cell type	**Ti-based SE alloy**	**Ti-Ni**	**Ti**	**Ti-6Al-4V**	
L929	Ti-18Zr-11Nb-3Sn	+			[13]
SaOS-2	Ti-18Zr-11Nb-3Sn	+			[13]
HUVEC	Ti-18Zr-11Nb-3Sn	O			[13]
COS7	Ti-24Nb-3Al		O		[14]
COS7	Ti-7Cr-3Sn		−		[14]
NIH 3T3	Ti-26Nb	+	O		[15]
hFOB 1.19	Ti-25Ta-25Nb			O	[16]
MC3T3-E1	Ti-23Nb-0.7Ta-2Zr-0.5N			+	[17]
RAW 264.7	Ti-23Nb-0.7Ta-2Zr-0.5N			O	[17]
MG63	Ti-23Nb-0.7Ta-2Zr-0.5N		O		[18]
MG63	Ti-19.1Nb-9.8Zr		−		[21]
MC3T3-E1	Ti-19Zr-10Nb-1Fe	+			[23]
V-79	Ti-3Mo-6Sn-5.2Zr		O		[24]
MC3T3-E1	Ti-3Mo-6Sn-5.2Zr		O		[24]
HUVEC	Ti-23Hf-3Mo-4Sn		O	O	[25]

+, superior cytocompatibility in comparison with control material; O, equivalent cytocompatibility in comparison with control material; −, inferior cytocompatibility in comparison with control material.

cytocompatibility with HUVEC in comparison with control substrates of the conventional pure Ti and Ti-6Al-4V.

The results of cytocompatibility evaluation of Ti-based superelastic alloys are summarized in Table 6.1. It is clearly noticed that Ti-based superelastic alloys exhibit excellent cytocompatibility at least comparable with Ti-Ni and Ti-6Al-4V in all cell lines investigated. In some cell lines, Ti-based superelastic alloys show superior cytocompatibility compared with Ti-Ni and Ti-6Al-4V. It is also seen that Ti-based superelastic alloys exhibit cytocompatibility almost equivalent to pure Ti.

6.3 ELECTROCHEMICAL AND CORROSION BEHAVIOR

Cytotoxicity of metallic materials strongly depends on the metallic ions released from metal surface and the toxicity of the ions. Therefore, the corrosion resistance is one of the important properties for biomedical applications. Electrochemical and corrosion properties have been investigated in some Ti-based shape memory alloys. Ijaz et al. [26] investigated

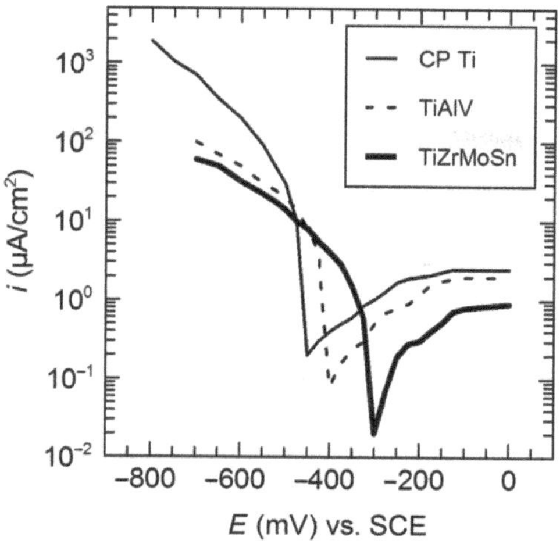

Fig. 6.6 Polarization curves for cp-Ti, Ti-6Al-4V, and (Ti-Zr)-1.5Mo-3Sn alloys in SBF at 310 K [26].

electrochemical properties of a (Ti-Zr)-1.5Mo-3Sn superelastic alloy in simulated body fluid (SBF) in comparison with pure Ti and Ti-6Al-4V. As seen from polarization curves in Fig. 6.6 [26], the corrosion density (i_{corr}) and corrosion rate of (Ti-Zr)-1.5Mo-3Sn are significantly lower than those of pure Ti and Ti-6Al-4V. Furthermore, the polarization resistance of (Ti-Zr)-1.5Mo-3Sn is estimated to be five times higher than those of pure Ti and Ti-6Al-4V, indicating that (Ti-Zr)-1.5Mo-3Sn exhibited the noblest electrochemical behavior among three alloys investigated. They claimed that a higher corrosion resistance of (Ti-Zr)-1.5Mo-3Sn is due to the most protective passive film on surface consisting of TiO_2, ZrO_2, MoO_2, and SnO_2 oxides. McMahon et al. [15] compared the corrosion resistance of Ti-26Nb and Ni-49.2Ti (Ti-50.8Ni) using 1 M NaCl or in Hank's balanced salt solution (HBSS) at 310 K. Ti-26Nb showed stable passivation at least 2000 mV in both solutions; however, Ni-49.2Ti exhibited a breakdown potential of 300–400 mV, indicating that Ti-26Nb is more corrosion-resistant than Ni-49.2Ti. Rosalbino et al. [27] investigated the corrosion behavior of Ti-Nb-Sn shape memory alloys in naturally aerated Ringer's solution at 310 K and compared with the conventional Ti-Ni shape memory alloy. The results showed that a Ti-18Nb-5Sn shape memory alloy has a greater tendency to form passive oxide film. Furthermore, they assessed that the corrosion protection ability of the oxide film on Ti-18Nb-5Sn is much stronger than that of Ti-Ni. Xue et al. [23] also reported that Ti-19Zr-

10Nb-1Fe exhibited twice lower value of the corrosion current density in in Hank's solution at 310 K than TiNi. They also assessed that the ion release rates for Ti, Zr, Nb, and Fe ions of Ti-19Zr-10Nb-1Fe are far lower than Ni ion release in Ti-Ni.

6.4 SUMMARY

As expected from the alloy composition, experimental results have confirmed that β Ti-based alloys are cytocompatible with various cell lines. Although the sensitivity of the cells for cytotoxicity differs depending on the kind of alloy and testing condition, β Ti-based superelastic alloys exhibit superior cytocompatibility compared with Ti-Ni and Ti-6Al-4V and are as biocompatible as pure Ti. The excellent cytocompatibility of β Ti-based superelastic alloys is strongly related to their strong corrosion resistance and reduced metallic ion release. Even though some metal ions are released from β Ti-based superelastic alloys, the ions are less toxic than Ni. Although the Ni release level of Ti-Ni is low enough to cause toxic reaction from cells, the risk of Ni still exists when they are used in an environment subject to friction and dynamic loading. From this point of view, Ni-free Ti-based superelastic alloys have a significant advantage over conventional Ti-Ni-based superelastic alloys.

REFERENCES

[1] T.W. Duerig, K.N. Melton, D. Stockel, C.M. Wayman, Engineering Aspects of Shape Memory Alloys, Elsevier Ltd., London, 1990
[2] K. Yamauchi, I. Ohkata, K. Tsuchiya, S. Miyazaki, Shape Memory and Superelastic Alloys-Technologies and Applications, Woodhead Publishing Limited, Cambridge, 2011.
[3] T. Yoneyama, S. Miyazaki, Shape Memory Alloys for Biomedical Applications, Woodhead Publishing Limited, Cambridge, 2009.
[4] F.W. Sunderman, Carcinogenicity of metal alloys in orthopedic prosthesis: clinical and experimental studies, Fundam. Appl. Toxicol. 13 (1989) 205–216.
[5] A. Yamamoto, R. Honma, M. Sumita, Cytotoxicity evaluation of 43 metal salts using murine fibroblasts and osteoblastic cells, J. Biomed. Mater. Res. 39 (1998) 331–340.
[6] A. Hartwig, Carcinogenicity of metal compounds: possible role of DNA repair inhibition, Toxicol. Lett. 102–103 (1998) 235–239.
[7] M. Assad, N. Lemieux, C. Rivard, L. Yahia, Comparative in vitro biocompatibility of nickel-titanium, pure nickel, pure titanium, and stainless steel: genotoxicity and atomic absorption evaluation, Bio-Med. Mater. Eng. 9 (1999) 1–12.
[8] A. Yamamoto, Y. Kohyama, T. Hanawa, Mutagenicity evaluation of forty-one metal salts by the umu test, J. Biomed. Mater. Res. 59 (2002) 176–183.
[9] D. Beyersmann, A. Hartwig, Carcinogenic metal compounds: recent insight into molecular and cellular mechanisms, Arch. Toxicol. 82 (2008) 493–512.

[10] A. Biesiekierski, J. Wang, M.A.H. Gepreel, C. Wen, A new look at biomedical Ti-based shape memory alloys, Acta Biomater. 7 (2012) 1661–1669.
[11] ISO 10993-5:2009(E) n.d., Biological Evaluation of Medical Devices—Part 5: Tests for in vitro Cytotoxicity, International Organization for Standardization, 2009.
[12] A. Yamamoto, R. Honma, A. Tanaka, M. Sumita, Generic tendency of metal salt cytotoxicity for six cell lines, J. Biomed. Mater. Res. 40 (1999) 396–403.
[13] J. Fu, A. Yamamoto, H.Y. Kim, H. Hosoda, S. Miyazaki, Novel Ti-base superelastic alloys with large recovery strain and excellent biocompatibility, Acta Biomater. 17 (2015) 56–67.
[14] H. Kanetaka, H. Hosoda, Y. Shimizu, T. Kudo, Y. Zhang, M. Kano, Y. Sano, S. Miyazaki, In vitro biocompatibility of Ni-free Ti-based shape memory alloys for biomedical applications, Mater. Trans. 51 (2010) 1944–1950.
[15] R.E. McMahon, J. Ma, S.V. Verkhoturov, D. Munoz-Pinto, I. Karaman, F. Rubitschek, H.J. Maier, M.S. Hahn, A comparative study of the cytotoxicity and corrosion resistance of nickel–titanium and titanium–niobium shape memory alloys, Acta Biomater. 8 (2012) 2863–2870.
[16] A. Cimpean, V. Mitran, C.M. Ciofrangeanu, B. Galateanu, E. Bertrand, D.M. Gordin, D. Iordachescu, T. Gloriant, Osteoblast cell behavior on the new beta-type Ti–25Ta–25Nb alloy, Mater. Sci. Eng. C 32 (2012) 1554–1563.
[17] R. Ion, D.M. Gordin, V. Mitran, P. Osiceanu, S. Dinescu, T. Gloriant, A. Cimpean, In vitro bio-functional performances of the novel superelastic beta-type Ti–23Nb–0.7 Ta–2Zr–0.5N alloy, Mater. Sci. Eng. C 33 (2014) 411–419.
[18] V. Mitran, C. Vasilescu, S.I. Drob, P. Osiceanu, J.M. Calderon-Moreno, M.C. Tabirca, D.M. Gordin, T. Gloriant, A. Cimpean, Biological behaviour and enhanced anticorrosive performance of the nitrided superelastic Ti-23Nb-0.7 Ta-2Zr-0.5N alloy, BioMed Res. Inter. 2015 (2015)261802.
[19] P. Castany, D.M. Gordin, S.I. Drob, C. Vasilescu, V. Mitran, A. Cimpean, T. Gloriant, Deformation mechanisms and biocompatibility of the superelastic Ti–23Nb–0.7 Ta–2Zr–0.5N alloy, Shap. Mem. Superelasticity 2 (2016) 18–28.
[20] K. Miura, N. Yamada, S. Hanada, T.K. Jung, E. Itoi, The bone tissue compatibility of a new Ti–Nb–Sn alloy with a low Young's modulus, Acta Biomater. 7 (2011) 2320–2326.
[21] M. Arciniegas, J. Pena, F.J. Gil, J.M. Manero, In vitro response of preosteoblastic MG63 cells on Ni-free Ti shape memory substrates, J. Biomed. Mater. Res. B 101B (2013) 709–720.
[22] M. Arciniegas, J.M. Manero, E. Espinar, J.M. Llamas, J.M. Barrera, F.J. Gil, New Ni-free superelastic alloy for orthodontic applications, Mater. Sci. Eng. C 33 (2013) 3325–3328.
[23] P. Xue, Y. Li, K. Li, D. Zhang, C. Zhou, Superelasticity, corrosion resistance and biocompatibility of the Ti–19Zr–10Nb–1Fe alloy, Mater. Sci. Eng. C 50 (2015) 179–186.
[24] S. Nunome, H. Kanetaka, T. Kudo, K. Endoh, H. Hosoda, K. Igarashi, In vitro evaluation of biocompatibility of Ti–Mo–Sn–Zr superelastic alloy, J. Biomater. App. 30 (2015) 119–130.
[25] R. Ion, S.I. Drob, M.F. Ijaz, C. Vasilescu, P. Osiceanu, D.M. Gordin, A. Cimpean, T. Gloriant, Surface characterization, corrosion resistance and in vitro biocompatibility of a new Ti-Hf-Mo-Sn alloy, Materials 9 (2016) 818.
[26] M.F. Ijaz, C. Vasilescu, S.I. Drob, P. Osiceanu, M. Marcu, H.Y. Kim, S. Miyazaki, D. M. Gordin, T. Gloriant, Electrochemical characterization of the superelastic (Ti-Zr)-Mo-Sn biomedical alloy displaying a large recovery strain, Mater. Corr. 68 (2017) 1220–1227.
[27] F. Rosalbino, D. Maccio, G. Scavino, A. Saccone, In vitro corrosion behaviour of Ti–Nb–Sn shape memory alloys in Ringer's physiological solution, J. Mater. Sci. Mater. Med. 23 (2012) 865–871.

CHAPTER 7

Fabrication and Characterization of Shape Memory Alloys

Contents

7.1 INTRODUCTION

Titanium and titanium alloys have been widely used in various fields, such as aerospace, chemical, automotive, and biomedical industries, and the manufacturing processes of titanium and titanium alloys have been well established [1]. Titanium and titanium alloys have been regarded as difficult to fabricate due to their high chemical affinity and reactivity with oxygen, nitrogen, and carbon. In order to prevent contamination, melting should be carried out in a high vacuum or in an atmosphere of high-purity inert gas. Although there have been successful attempts to develop Ni-free Ti-based shape memory alloys, until recently, there have been limited commercial alloys available. On the other hand, Ti-Ni-based shape memory alloys have been successfully used in various fields [2–4]. The manufacturing processes of Ti-Ni-based shape memory alloys such as melting, hot working, cold working, heat treatment, machining, and surface treatment have been well established [3–11]. One of the most important issues in fabrication process is to obtain chemically homogeneous and uniform structure since the martensitic transformation temperatures and shape memory properties are highly dependent on the composition variation and microstructure. In this chapter, several issues on the fabrication process of commercial Ti-Ni shape memory alloys are overviewed. The fabrication process and thermomechanical treatment of recently developed Ni-free Ti-based shape memory alloys are also

Ni-free Ti-based Shape Memory Alloys
https://doi.org/10.1016/B978-0-12-809401-3.00007-0

summarized. The characterization methods for basic properties of shape memory alloys are briefly described.

7.2 FABRICATION PROCESS OF Ti-Ni SHAPE MEMORY ALLOYS

There have been no commercially available Ni-free Ti-based shape memory alloys until recently; therefore, the fabrication process of Ti-Ni-based shape memory alloys is described first. The most popular industrial method for the fabrication of Ti-Ni shape memory alloy is ingot metallurgy. The fabrication process is divided into the following steps (Fig. 7.1): ingot melting; hot working; cold working; heat treatment to achieve shape memory effect; and subsequent manufacturing process such as cutting, welding, and surface treatment. Transformation temperatures of Ti-Ni shape memory alloys are highly sensitive to alloy composition. Therefore, it is essential to avoid composition variation or contamination during fabrication process of shape memory alloys. One of the key issues in the fabrication process is to control the interstitial impurity elements such as oxygen and carbon [5–11]. Both of the quality of raw materials and melting method are important to fabricate ingots of high purity and homogeneity.

The most common method for producing Ti-Ni ingots is vacuum melting such as vacuum induction melting (VIM) and vacuum arc remelting (VAR) because of the high reactivity of titanium with oxygen. In the case of VIM, the raw materials are charged into a graphite crucible or a calcium oxide crucible and then melted by high-frequency induction heating. VIM

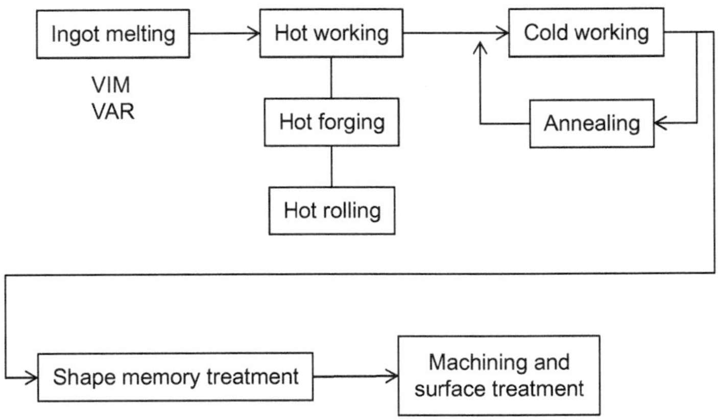

Fig. 7.1 Typical fabrication process of Ti-Ni shape memory alloys.

allows a good homogenization of the melt; therefore, the ingots produced by VIM are generally homogeneous and uniform. One possible concern is contamination due to reaction with the crucible. Particularly, the contamination of carbon is a major issue when the graphite crucible is used because carbon is easily dissolved into molten TiNi and forms TiC during solidification. The formation of TiC decreases the Ti content of the ingots, resulting in the decrease of the martensitic transformation temperatures. On the other hand, the use of oxide crucibles such as CaO, MgO, and Al_2O_3 has a greater risk of the formation of oxide-type compounds, Ti_2NiO_x or Ti_4Ni_2O. These compounds also affect the transformation temperatures and cause detrimental effects on the workability and mechanical properties. ASTM F2063 specifies the maximum acceptable carbon and oxygen contents of 500 wt% ppm for the Ti-Ni shape memory alloy ingots for medical devices and surgical implants [12]. In the case of VAR, materials are melted by a plasma arc between the electrode and the base of a water-cooled copper mold in vacuum. Generally, a cylinder-type consumable electrode is used in VAR process. VAR can eliminate the risk of carbon contamination. The impurity levels of the ingot are dependent on the quality of raw materials and vacuum condition. One major disadvantage of VAR is that the ingots made by VAR are less homogeneous in composition and less uniform in microstructure when compared with VIM. The chemical composition of an ingot varies from top to bottom because only a part of the ingot is molten and the molten part is contacted with the solidified part. In order to improve the homogeneity and uniformity, the ingot is required to be remelted multiple times. A combination of two processes (VIM/VAR) has also been applied to produce Ti-Ni ingots. Other melting techniques, such as electron beam melting (EBM) [13–15] and induction skull melting (ISM) [16], have been applied to obtain high-purity and homogeneous ingots.

After melting, conventional thermomechanical process is applied to fabricate final product form such as wire, plate, and tube. At first, the as-cast ingots are hot forged or hot-rolled to break down the cast structure. This hot working is generally carried out in the temperature range of 973–1223 K. Cold working, e.g., cold rolling and cold drawing, is performed to obtain the final form of products such as wire, plate, and tube. Poor cold workability of Ti-Ni shape memory alloys allows reductions less than 50%, and then, the alloys need to be softened by intermediate annealing in the temperature range of 973–1073 K for further cold working. It is noted that accurate dimensional control during cold working is difficult because of strong springback effect of shape memory alloys. Multiple cycles of cold working and intermediate annealing are required to achieve thin wire or plate. Typically, final cold working ratio is controlled to be 20%–50%. After the final cold working, heat treatment is applied not only to memorize a desired shape but also to adjust transformation temperatures.

7.3 FABRICATION PROCESS OF Ni-free Ti-based SHAPE MEMORY ALLOYS

Melting temperature is much higher for Ni-free Ti-based shape memory alloys compared with Ti-Ni alloys because they contain β-phase-stabilizing alloying elements with high melting points such as Nb, Ta, and Mo. Therefore, it seems much difficult to apply VIM for the fabrication of Ni-free Ti-based shape memory alloy ingots. Another important issue is the control of interstitial elements, particularly oxygen, because the solubility limit of interstitial elements in β Ti alloys is high and the interstitial elements strongly affect the martensitic transformation behavior. The solubility limit of oxygen is about 8 and 33 at.% in β phase and α phase of Ti, respectively. The cold workability of β Ti alloys is excellent compared with Ti-Ni alloys because of their low work-hardening rate. Some β Ti alloys can be cold worked more than 99% without intermediate annealing.

VAR has been a principal method for the fabrication of Ti alloy ingots since the 1950s. Cold hearth melting (CHM) has become another major melting technique for Ti and its alloys since the 1980s [1]. In CHM process, the materials are melted in a water-cooled vessel using electron beam or plasma arc. CHM has an advantage of removing inclusions both high-density inclusions like WC and low-density inclusions like TiN due to a solid skull layer formed between the molten part and copper vessel. Heavy inclusions sink into the skull, and low-density inclusions are dissolved into the melt before flowing into a crucible, resulting in high-quality ingots.

So far, only laboratory-scale research has been carried out for the Ni-free Ti-based shape memory alloys. Typically small sizes of ingots 10–100 g have been fabricated by Ar arc melting using nonconsumable tungsten electrode on water-cooled copper hearth. The chamber is evacuated and then flushed using high-purity argon gas. The ingots are required to be melted multiple times, typically five or six times, being flipped over after each melting step in order to improve homogeneity in composition. Many Ti-based shape memory alloys, such as Ti-Nb-based [17–23], Ti-Ta-based [24–27], Ti-Zr-based [28–31], and Ti-Mo-based [32–35] alloys, have been successfully fabricated using the Ar arc melting method. Kent et al. [36] fabricated a 25 kg ingot of a

Ti-25Nb-3Zr-3Mo-2Sn (wt%) alloy using arc melting. They used Nb-47 wt% Ti binary alloy as a master alloy instead of pure Nb to improve homogeneity. Cold crucible levitation melting (CCLM) has been also applied to fabricate Ti-based shape memory alloys, e.g., Ti-25Ta-25Nb [37], Ti-26Nb [38], Ti-20Nb-6Zr [38], Ti-23Nb-0.7Ta-2Zr-0.5N [39], Ti-24Nb-0.5O [40], Ti-24Nb-0.5N [40], and Ti-23Nb-3Mo-4Sn [41]. CCLM is particularly useful to fabricate Ti alloys containing high-melting-point elements because it allows uniform composition without contamination due to a strong stirring ability caused by an electromagnetic force. Conventional thermomechanical processes have been applied to make thin plates and wires for characterization. Fig. 7.2 shows a typical processing route of Ni-free Ti-based shape memory alloys. Homogenization treatment at above β transus temperature is commonly applied to remove the microsegregation and to improve the chemical homogeneity. Hot working, i.e., hot forging or hot rolling, is carried out for improving uniformity and reducing the size of the billet. After hot working, cold rolling or drawing is carried out to obtain specimens. Cold workability of β Ti alloys is excellent: some alloys can be cold-rolled up to 99.9% reduction in thickness without intermediate annealing. Therefore, in many laboratory-scale studies, hot working process has been skipped. After cold working, solution treatment or annealing is performed at near-β-transus temperature to optimize the microstructure. In some alloys, following aging treatment is effective to improve the shape

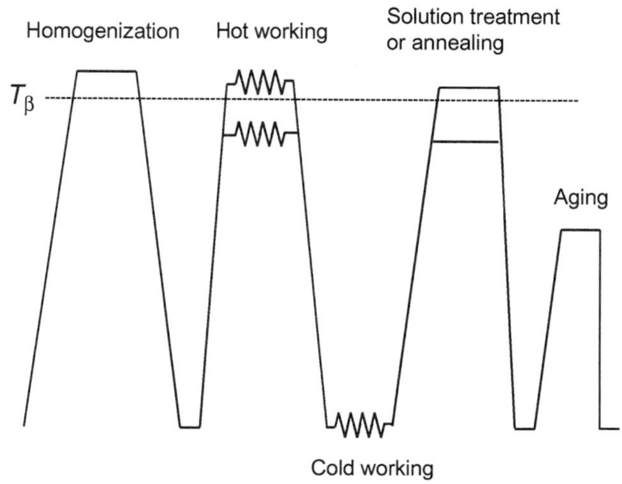

Fig. 7.2 Typical processing route of Ni-free Ti-based shape memory alloys.

memory effect. The details about the effect of thermomechanical treatment on the microstructure and shape memory properties are given in Chapter 4.

7.4 CHARACTERIZATION OF SHAPE MEMORY ALLOYS

One of the important parameters governing the shape memory properties is the martensitic transformation temperature and the reverse transformation temperature. The principal method used in these days to determine the transformation temperatures is differential scanning calorimetry (DSC). Fig. 7.3 shows an example of DSC curves showing martensitic transformation. From the exothermic peak upon cooling, the martensitic transformation start temperature (M_s), martensitic transformation finish temperature (M_f), and martensitic transformation peak temperature (M^*) are determined. Also, the reverse transformation start temperature (A_s), reverse transformation finish temperature (A_f), and reverse transformation peak temperature (A^*) can be determined from the corresponding endothermic peak upon heating. Other techniques, e.g., metallographic, electric-resistance, and thermal-arrest techniques, have been applied to detect the transformation temperatures. For example, in the 1950s–1960s, martensitic transformation start temperature in several Ti-Nb binary alloys has been determined using the thermal-arrest technique [42–44].

Fundamental properties and mechanisms of shape memory effect and superelasticity are well documented in literature [3, 4, 45–50]. Here, we

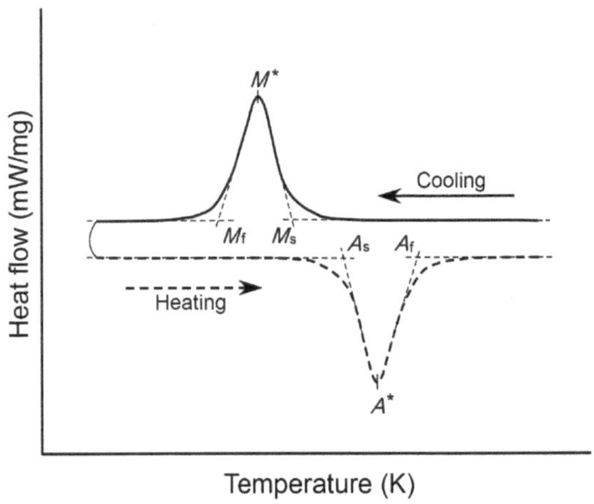

Fig. 7.3 Schematic DSC curves of shape memory alloys showing martensitic transformation and corresponding reverse transformation.

briefly describe some of the basic features of the shape memory effect and superelasticity. Shape memory properties are characterized by stress–strain-temperature relationships. Fig. 7.4A shows a schematic representation of shape memory effect. Below M_f, the specimen is fully transformed to martensite phase consisting of multivariants with twinning relationship. Applying stress causes detwinning in the twinned martensite phase by the movement of twin boundary at a certain critical stress noted as σ_T. After unloading, the detwinned martensite phase is remained, i.e., the apparent plastic strain is remained, because the martensite phase is stable at M_f. Upon heating, the recovery of the residual strain occurs due to the reverse transformation starting at A_s and finishing at A_f. In other words, the reverse

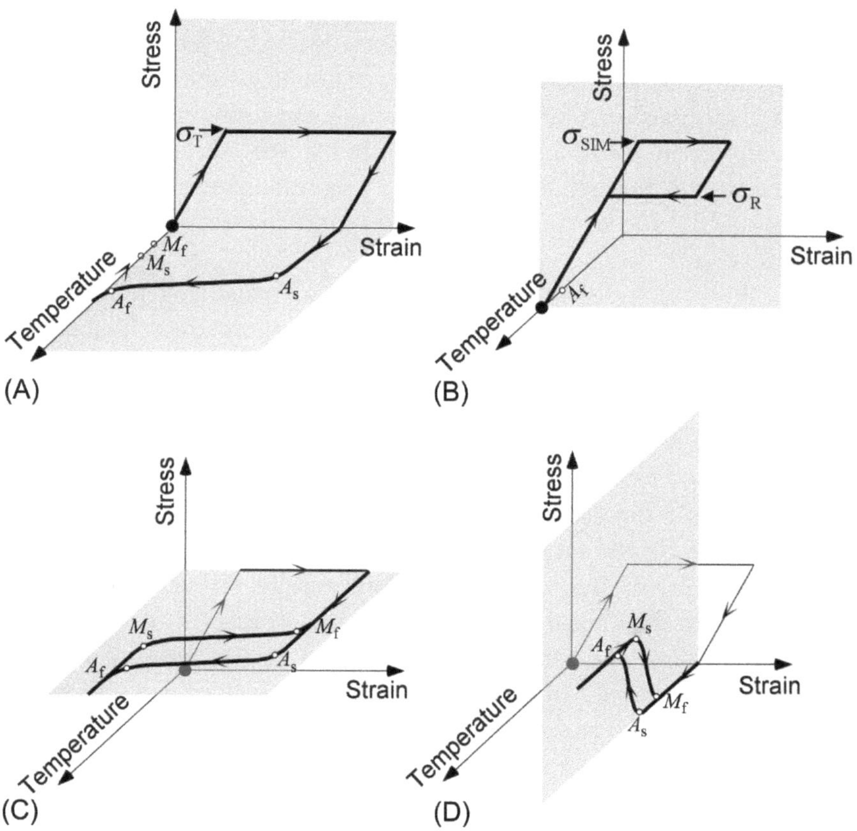

Fig. 7.4 Schematic illustrations describing stress-strain-temperature relationships of shape memory alloys: (A) shape memory effect cycle, (B) superelastic cycle, (C) strain-temperature relationship at a constant stress, and (D) stress-temperature relationship at a constant strain.

transformation temperatures, A_s and A_f, are easily determined by recording the shape change upon heating the specimen deformed below M_f. ASTM F2082 [51] specifies the method for measuring the active reverse transformation finish temperature (active A_f) using bend-free recovery test. Fig. 7.4B shows a schematic representation of superelasticity. When the stress is applied above A_f, the stress-induced martensitic transformation occurs if the onset stress for inducing martensitic transformation (σ_{SIM}) is lower than the stress for plastic deformation. Upon unloading, the strain recovers at the onset stress for reverse transformation (σ_R). Fig. 7.4C shows a strain-temperature cycle at a constant applied stress. Similarly to Fig. 7.4A, the strain recovery occurs upon heating the deformed specimen. Upon cooling the specimen, the specimen elongates again due to the martensitic transformation, because the martensitic transformation occurs at a constant applied stress. Fig. 7.4D shows a stress-temperature cycle at a fixed strain. Upon heating, the stress starts to increase at A_s because the shape recovery is suppressed. The stress is relieved again below M_s due to the martensitic transformation upon cooling. It should be noted that transformation temperatures determined by these cycles may differ from the values measured by DSC due to the effect of stress and strain on the martensitic transformation.

The effect of stress on the martensitic transformation temperatures is explained by the Clausius-Clapeyron relationship of the stress-induced martensitic transformation as shown in Fig. 7.5. The stress required for inducing martensitic transformation increases linearly with increasing temperature according to the following equation: $d\sigma_{SIM}/dT = -\Delta H/T \cdot \varepsilon$, where σ_{SIM} is the onset stress for inducing martensitic transformation, ε is transformation strain, and ΔH is transformation enthalpy per unit volume. Similar relationship also holds for the reverse transformation. Accordingly, both σ_{SIM} and σ_R increase with increasing test temperature as shown in Fig. 7.5B. Similarly, the strain-temperature cycle curve obtained at a higher stress σ_2 shows higher transformation temperatures as shown in Fig. 7.5C.

For the applications of superelastic alloys, not only the stress levels of σ_{SIM} and σ_R but also recoverable strain is important. Typically, the superelastic properties are evaluated by the loading-unloading cycle tests. Fig. 7.6 shows an example of the stress-strain curves obtained by a strain increment loading-unloading cycle test. The applied strain recovers almost completely upon unloading at first several cycles. With increasing applied strain, the shape recovery becomes incomplete, and the residual strain increases. The residual strain drastically increases when the deformation proceeds beyond

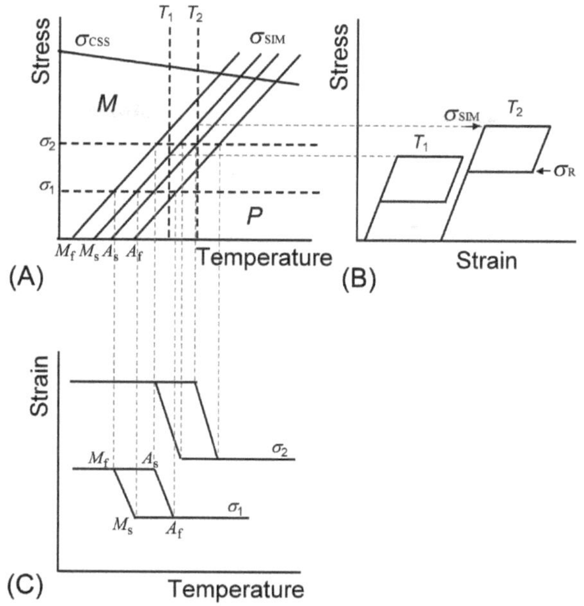

Fig. 7.5 Schematic illustrations describing the stress on the shape memory properties. (A) Clausius-Clapeyron relationship of the stress-induced martensitic transformation, (B) the effect of test temperature on superelastic behavior, and (C) the effect of applied stress on the strain-temperature curve.

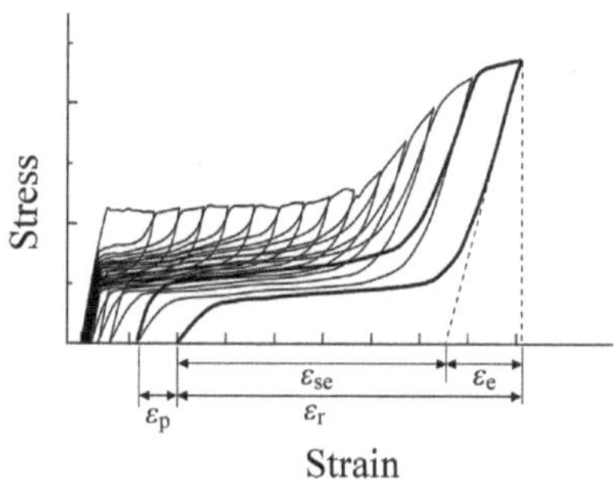

Fig. 7.6 Schematic stress-strain curves obtained from a strain increment loading-unloading cycle test.

stress plateau region. At each cycle, four types of strains, superelastic recovery strain (ε_{se}), elastic strain (ε_e), recovery strain (ε_r), and residual strain (ε_p), can be derived from the stress-strain curve obtained at each cycle. The maximum recovery strain is affected by various factors such as lattice distortion strain, texture, loading direction, critical stress for slip deformation, and test temperature.

7.5 SUMMARY

Comparing with Ti-Ni shape memory alloys, it is pointed out that the following differences are to be noted for the fabrication process of Ni-free Ti-based shape memory alloys: (1) melting temperature, (2) solubility of interstitial elements, and (3) cold workability. Titanium has a strong chemical affinity to oxygen so that the melting process for titanium alloys is performed in vacuum or inert gas atmosphere. Two common methods for producing Ti-Ni ingots are VIM and VAR. On the other hand, only VAR has been a principal method for the fabrication of β Ti alloy ingots because of high melting temperature. In Ti-Ni alloys, the formation of oxide-type compounds is a major concern during melting because they reduce the transformation temperatures and cause detrimental effects on the workability and mechanical properties. Unlike Ti-Ni alloys, oxygen has a large solubility in β Ti alloys; therefore, a large amount of oxygen is present as interstitial atoms. Poor cold workability is one of the major causes for high processing cost of Ti-Ni. On the other hand, β Ti alloys exhibit excellent cold workability. Although most of researches have been performed in laboratory scale, the results have shown that Ni-free Ti-based shape memory alloys can be fabricated and processed using conventional techniques of melting and thermomechanical treatment established in Ti industries.

REFERENCES

[1] G. Lütjering, J.C. Williams, Titanium, Springer-Verlag Berlin Heidelberg, 2007.
[2] T.W. Duerig, K.N. Melton, D. Stockel, C.M. Wayman, Engineering Aspects of Shape Memory Alloys, Elsevier Ltd., 1990
[3] K. Yamauchi, I. Ohkata, K. Tsuchiya, S. Miyazaki, Shape Memory and Superelastic Alloys-Technologies and Applications, Woodhead Publishing Limited, 2011.
[4] T. Yoneyama, S. Miyazaki, Shape Memory Alloys for Biomedical Applications, Woodhead Publishing Limited, 2009.
[5] Y. Suzuki, K. Otsuka, C.M. Wayman (Eds.), Fabrication of Shape Memory Alloys (Shape Memory Materials), Cambridge University Press, 1998.
[6] S.M. Russell, S.M. Russell, A.R. Pelton (Eds.), Nitinol melting and fabrication, Proceedings of the International Conference on Shape Memory and Superelastic Technologies SMST-2000, SMST, 2001, pp. 1–10.
[7] J. Frenzel, Z. Zhang, K. Neuking, G. Eggeler, High quality vacuum induction melting of small quantities of NiTi shape memory alloys in graphite crucibles, J. Alloys Compd. 28 (2004) 214–223.

[8] T. Nakahata, Industrial processing of titanium-nickel (Ti-Ni) shape memory alloys (SMAs) to achieve key properties, in: K. Yamauchi, I. Ohkata, K. Tsuchiya, S. Miyazaki (Eds.), Shape Memory and Superelastic Alloys-Technologies and Applications, Woodhead Publishing Limited, 2011, pp. 53–62.
[9] M.H. Elahinia, M. Hashemi, M. Tabesh, S.B. Bhaduri, Manufacturing and processing of NiTi implants: a review, Prog. Mater. Sci. 57 (2012) 911–946.
[10] N. Sharma, K.K. Jangra, T. Raj, Fabrication of NiTi alloy: a review, Proc. Inst. Mech. Eng. L: J. Med. Des. Appl. 232 (2015) 250–269.
[11] M.H. Elahinia, Shape Memory Alloy Actuators: Design, Fabrication, and Experimental Evaluation, John Wiley & Sons, 2016.
[12] ASTM F 2063-05 Standard Specification for Wrought Nickel-Titanium Shape Memory Alloys for Medical Devices and Surgical Implants, ASTM International, 2005.
[13] J. Otubo, O.D. Rigo, C.D.M. Neto, M.J. Kaufman, P.R. Mei, Scale up of NiTi shape memory alloy production by EBM, J. Phys. IV 112 (2003) 873–876.
[14] J. Otubo, O.D. Rigo, C.D.M. Neto, M.J. Kaufman, P.R. Mei, Low carbon content NiTi shape memory alloy produced by electron beam melting, J. Mater. Res. 7 (2004) 263–267.
[15] J. Otubo, O.D. Rigo, C.D.M. Neto, P.R. Mei, The effects of vacuum induction melting and electron beam melting techniques on the purity of NiTi shape memory alloys, Mater. Sci. Eng. A 438–440 (2006) 678–682.
[16] G.M. Kramer, A comparison of chemistry and inclusion distribution and morphology versus melting method of NiTi alloys, J. Mater. Eng. Perform. 18 (2009) 479–483.
[17] H.Y. Kim, Y. Ikehara, J.I. Kim, H. Hosoda, S. Miyazaki, Martensitic transformation, shape memory effect and superelasticity of Ti–Nb binary alloys, Acta Mater. 54 (2006) 2419–2429.
[18] E. Takahashi, T. Sakurai, S. Watanabe, N. Masahashi, S. Hanada, Effect of heat treatment and Sn content on superelasticity in biocompatible TiNbSn alloys, Mater. Trans. 43 (2002) 2978–2983.
[19] Y. Fukui, T. Inamura, H. Hosoda, K. Wakashima, S. Miyazaki, Mechanical properties of a Ti–Nb–Al shape memory alloy, Mater. Trans. 45 (2004) 1077–1082.
[20] Y.L. Hao, S.J. Li, S.Y. Sun, R. Yang, Effect of Zr and Sn on Young's modulus and superelasticity of Ti–Nb-based alloys, Mater. Sci. Eng. A 441 (2006) 112–118.
[21] D.C. Zhang, S. Yang, M. Wei, Y.F. Mao, C.G. Tan, J.G. Lin, Effect of Sn addition on the microstructure and superelasticity in Ti–Nb–Mo–Sn alloys, J. Mech. Behav. Biomed. Mater. 13 (2012) 156–165.
[22] Y. Zhu, L. Wang, M. Wang, Z. Liu, J. Qin, D. Zhang, W. Lu, Superelastic and shape memory properties of TixNb3Zr2Ta alloys, J. Mech. Behav. Biomed. Mater. 12 (2012) 151–159.
[23] E.G. Obbard, Y.L. Hao, R.J. Talling, S.J. Li, Y.W. Zhang, D. Dye, R. Yang, The effect of oxygen on α'' martensite and superelasticity in Ti–24Nb–4Zr–8Sn, Acta Mater. 59 (2011) 112–125.
[24] P.J.S. Buenconsejo, H.Y. Kim, S. Miyazaki, Shape memory behavior of Ti–Ta and its potential as a high-temperature shape memory alloy, Acta Mater. 57 (2009) 1068–1077.
[25] P.J.S. Buenconsejo, H.Y. Kim, S. Miyazaki, Effect of ternary alloying elements on the shape memory behavior of Ti–Ta alloys, Acta Mater. 57 (2009) 2509–2515.
[26] H.Y. Kim, T. Fukushima, P.J.S. Buenconsejo, T.H. Nam, S. Miyazaki, Martensitic transformation and shape memory properties of Ti–Ta–Sn high temperature shape memory alloys, Mater. Sci. Eng. 528 (2011) 7238–7246.
[27] P.J.S. Buenconsejo, H.Y. Kim, S. Miyazaki, Novel beta-TiTaAl alloys with excellent cold workability and a stable high-temperature shape memory effect, Scr. Mater. 64 (2011) 1114–1117.

[28] Y. Cui, Y. Li, K. Luo, H.B. Xu, Microstructure and shape memory effect of Ti–20Zr–10Nb alloy, Mater. Sci. Eng. A 527 (2010) 652–656.
[29] J. Fu, A. Yamamoto, H.Y. Kim, H. Hosoda, S. Miyazaki, Novel Ti-base superelastic alloys with large recovery strain and excellent biocompatibility, Acta Biomater. 17 (2015) 56–67.
[30] L.L. Pavón, H.Y. Kim, H. Hosoda, S. Miyazaki, Effect of Nb content and heat treatment temperature on superelastic properties of Ti–24Zr–(8–12)Nb–2Sn alloys, Scr. Mater. 95 (2015) 46–49.
[31] M.F. Ijaz, H.Y. Kim, H. Hosoda, S. Miyazaki, Superelastic properties of biomedical (Ti–Zr)–Mo–Sn alloys, Mater. Sci. Eng. C 48 (2014) 11–20.
[32] T. Maeshima, S. Ushimaru, K. Yamauchi, M. Nishida, Effects of Sn content and aging conditions on superelasticity in biomedical Ti–Mo–Sn alloys, Mater. Trans. 47 (2006) 513–517.
[33] H.Y. Kim, Y. Ohmatsu, J.I. Kim, H. Hosoda, S. Miyazaki, Mechanical properties and shape memory behavior of Ti–Mo–Ga alloys, Mater. Trans. 45 (2004) 1090–1095.
[34] T. Zhou, M. Aindow, S.P. Alpay, M.J. Blackburn, M.H. Wu, Pseudo-elastic deformation behavior in a Ti/Mo-based alloy, Scr. Mater. 50 (2004) 343–348.
[35] H.Y. Kim, Y. Ohmatsu, J.I. Kim, T. Inamura, H. Hosoda, S. Miyazaki, Effect of Nb addition on shape memory behavior of Ti–Mo–Ga alloys, Mater. Trans. 47 (2006) 518–522.
[36] D. Kent, G. Wang, Z. Yu, M.S. Dargusch, Pseudoelastic behaviour of a β Ti–25Nb–3Zr–3Mo–2Sn alloy, Mater. Sci. Eng. A 527 (2010) 2246–2252.
[37] E. Bertrand, T. Gloriant, D.M. Gordin, E. Vasilescu, P. Drob, C. Vasilescu, S.I. Drob, Synthesis and characterization of a new superelastic Ti–25Ta–25Nb biomedical alloy, J. Mech. Behav. Biomed. Mater. 3 (2010) 559–564.
[38] F. Sun, Y.L. Hao, S. Nowak, T. Gloriant, P. Laheurte, F. Prima, A thermal-mechanical treatment to improve the superelastic performances of biomedical Ti–26Nb and Ti–20Nb–6Zr(at.%) alloys, J. Mech. Behav. Biomed. Mater. 4 (2011) 1864–1872.
[39] P. Castany, D.M. Gordin, S.I. Drob, C. Vasilescu, V. Mitran, A. Cimpean, T. Gloriant, Deformation mechanisms and biocompatibility of the superelastic Ti–23Nb–0.7Ta–2Zr–0.5N alloy, Shap. Mem. Superelasticity 2 (2016) 18–28.
[40] A. Ramarolahy, P. Castany, F. Prima, P. Laheurte, I. Péron, T. Gloriant, Microstructure and mechanical behavior of superelastic Ti–24Nb–0.5O and Ti–24Nb–0.5N biomedical alloys, J. Mech. Behav. Biomed. Mater. 9 (2012) 83–90.
[41] M.F. Ijaz, D. Laillé, L. Héraud, D.M. Gordin, P. Castany, T. Gloriant, Design of a novel superelastic Ti–23Hf–3Mo–4Sn biomedical alloy combining low modulus, high strength and large recovery strain, Mater. Lett. 177 (2016) 39–41.
[42] P. Duwez, Effect of rate of cooling on the alpha–beta transformation in titanium and titanium–molybdenum alloys, JOM 3 (1951) 765–771.
[43] P. Duwez, The martensitic transformation temperature in titanium binary alloys, Trans. ASM 45 (1953) 934–940.
[44] H. Kaneko, Y.C. Huang, Allotropic transformation characteristics of titanium alloys during continuous cooling, J. Jpn. Inst. Met. 27 (1963) 387–393.
[45] S. Miyazaki, S. Kimura, F. Takei, T. Miura, K. Otsuka, Y. Suzuki, Shape memory effect and pseudoelasticity in a Ti–Ni single crystal, Scr. Metall. 17 (1983) 1057–1062.
[46] S. Miyazaki, K. Otsuka, Development of shape memory alloys, ISIJ Int. 29 (1989) 353–377.
[47] S. Miyazaki, K. Otsuka, C.M. Wayman, The shape memory mechanism associated with the martensitic transformation in Ti–Ni alloys: part II: variants coalescence and shape recovery, Acta Metall. 37 (1989) 1885–1890.

[48] S. Miyazaki, Development and characterization of shape memory alloys, in: Shape Memory Alloys. International Centre for Mechanical Sciences (Courses and Lectures), Vol. 351, Springer, Vienna, 1996.
[49] K. Otsuka, C.M. Wayman, Shape Memory Materials, Cambridge University Press, 1999.
[50] S. Miyazaki, Y.Q. Fu, W.M. Huang, Thin Film Shape Memory Alloys: Fundamentals and Device Applications, Cambridge University Press, 2009.
[51] ASTM F2082 Standard Test Method for Determination of Transformation Temperature of Nickel-Titanium Shape Memory Alloys by Bend and Free Recovery, ASTM International, 2016.

INDEX

Note: Page numbers followed by *f* indicate figures, and *t* indicate tables.

V

Y